The AI Human

Navigating a Transformed World

Dr. Blaine Fisher

To Mom and Dad, for investing in my education and dreams, and for being my one constant in an exponentially changing world. In a book about artificial intelligence, you remain my favorite example of the genuine kind.

Contents

Preface

This book started in rooms where people were trying to figure something out. Classrooms, mostly. Conference rooms. Coffee shops near campus where the Wi-Fi was decent and the questions were hard. How do we work well when the machines can write? How do we teach when the assignments we've been giving for thirty years can be finished in thirty seconds? How do we keep ourselves in the work?

I wrote chapters as field notes. Tested them on Tuesdays. Revised them after workshops where someone would raise a hand and say, wait, but what about this. The ideas that survived are the ones that held up under ordinary pressure: a deck due at noon, a policy draft that actually has to ship, a student who looks you in the eye and asks if any of this still matters.

Often I sat with my laptop open, watching other laptops glow in a half-circle around a table. There was always coffee. There was always someone who hadn't slept enough. And there was always this hum underneath the conversation, the awareness that we were living through something large and fast, and none of us were entirely sure what to do about it.

You won't find proofs here. You'll find something closer to a field guide. Human-centered. Practice-driven. The focus is less on what a model is and more on what people do with it. When a system helps you write, how do you keep your voice? When a copilot speeds you up, how do you keep your standards? When an agent can watch your inbox or summarize your meetings, how do you decide what stays human, what moves to the machine, and what has to be checked twice?

A few ideas run through everything that follows.

First: AI is infrastructure. Like electricity, it's becoming a general purpose technology that flows under much of what we do. When you see it that way, you stop chasing features and start building habits. You ask better questions. You verify with intent. You document how decisions get made. You keep a person responsible for what happens.

Second: the center of gravity is shifting. I call it the Great Promotion. As more tasks become assistable, our value moves up the stack, toward scoping, reviewing, coaching, stewarding the systems that now do the work we used to do by hand. That promotion is an opportunity, but only if we take it. It requires new literacy. Orchestration skills. Calm evaluation when the clock is running. A clear sense of where human judgment must stay in the loop.

Third: creativity and scholarship are changing in kind, not only in speed. Writing with AI can make mediocre faster, or it can make thoughtful braver. The difference is process. Draft out loud. Verify claims. Track sources. Separate search from synthesis. Use the machine to expand your range, not to erase your fingerprints.

Fourth: education isn't ending. It's upgrading. The so-called homework apocalypse is a design problem. When assignments measure recall, they get automated. When assignments ask students to direct a system, critique its output, and defend their choices, learning deepens. I share patterns that have worked in my courses: oral checks, transparent rubrics, prompts that reward original thinking.

Fifth: wonder fades. It always does. After the first wow comes the real work. Policy. Energy costs. Governance. Attention. This book steps back to those questions too, because responsible adoption isn't a slogan. It's a habit. And habits need scaffolding.

How to use this: dip in where the need is urgent, then circle back for context. If you manage teams, start with the chapters on workflow design, agents, and the Great Promotion. If you teach, pair the education chapters with the ones on writing and verification. If you lead creative work, try the sections on image generation, evaluation, protecting a house style. Each chapter ends with a small set of questions. Not homework. Prompts you can sit with, journal on, or bring into conversation as you decide what role you want AI to play in your life and work.

A note on voice. I write as a practitioner who teaches and a teacher who practices. I have ADHD. AI has been a powerful support in turning scattered ideas into shipped work. That shows up in the techniques I share, not just patterns and prompts but reflection questions that keep a human voice in the loop. If something here fits your brain, steal it. If it doesn't, revise it until it does.

If you're picking this up with optimism, welcome. If you're picking it up with hesitation, welcome. The point isn't to be first. It's to be good. Start small. Measure honestly. Keep people at the center. Let the models do what they're good at: prediction, pattern, speed. Let humans do what we must protect: meaning, ethics, taste, responsibility.

Thank you for reading. Challenge what follows. Mark the margins. Send me what you build. If this book does its job, you'll leave with a sharper mental model, a sturdier playbook, and more confidence that you can navigate, and improve, the systems that now shape our days.

Dr. Blaine Fisher

New Orleans, September 2025

***

How AI Helped, What Stayed Human

I wrote this book, start to finish, then used AI as a support tool. I brainstormed out loud, asked for outlines, and pressure-tested gaps. Every idea, argument, and first draft is mine. AI, including Grammarly, helped with structure, clarity, and final polish, more like autocorrect than ghostwriter. I verified facts, kept sources, and made the calls. A human, me, stayed responsible for what is claimed and how it is said. First drafts were messy. That is normal. Later drafts made it look like I knew what I was doing all along.

✓ Human in the loop. ✓ Facts checked. ✓ Style mine. ✓ Errors mine to own.

INTRODUCTION

You feel it in small ways first. The pocket tap when your phone isn't there. That little panic, physical and immediate. Or the playlist that somehow knows you need something slow at 4 p.m. on a Thursday. The email that surfaces exactly when you need it, not because you went looking but because something under the surface went looking for you.

These aren't party tricks anymore. They're signals. We don't simply use devices. We live through them. Artificial intelligence slipped from headline to habitat somewhere between 2020 and now, from spectacle to the quiet layer underneath. It's choosing your next song. Surfacing emails. Predicting meetings you forgot to schedule. In practical terms, AI now functions like any other utility. Electricity. Running water. A current beneath modern work and play.

The urgency isn't alarm. It's agency. We already opted in. The question is what we do now that we're here.

For me, the moment I understood this wasn't in a lab. It was in a library corner, late afternoon light coming through tall windows, watching a student ask a chatbot about Greek tragedies. His eyes lit up, not because he got the right answer but because the machine's reply sparked a question he hadn't thought to ask. That flicker. That's what keeps me interested in these tools. Not what they replace. What they make possible.

The claim, and why it matters

Before we get tactical, we need a clear stance.

This book makes a simple claim: teams that learn to design human and model collaboration, then measure results with care, will outperform those that chase tools without changing habits.

The change ahead isn't binary. It's supervisory. As more tasks become assistable, we get promoted. Not always up, but always toward roles that scope, review, and steward systems instead of executing tasks. Managers of machines. Stewards of consequences. Call it what you want, but the work is different now.

Two tensions run through what follows.

First, the productivity paradox. Powerful tools don't guarantee gains if we mismeasure value, misdesign workflows, or mistrain teams. I've watched entire departments buy the best software and somehow get slower.

Second, the experience paradox. Constant connection can exhaust attention and fray meaning unless we build guardrails and habits on purpose. The always-on life isn't sustainable. It never was.

The stakes are practical. Better classes. Cleaner drafts. Safer workflows. Choices you can defend in public. AI literacy should sit with reading, writing, and mathematics as a foundational skill, not a nice-to-have for tech people. My fear isn't that AI becomes too smart. It's that too few people understand it well enough to use it effectively, efficiently, and ethically. Ignorance at scale is expensive.

Who this is for, and what isn't here

Readers arrive with different roles, but the pressure feels similar.

Managers ship work without dropping standards. Faculty assess learning when the old tests don't work anymore. Students must show thinking, not just output. Creators protect voice and ethics when the tools can mimic both. Researchers keep a clean trail from claim to source in an age of infinite generated text.

This book meets you at that shared pressure point.

It focuses on habits, roles, and standards that survive version changes. Tools shift. Principles travel. You won't find an API cookbook here. No leaderboard of models. No vendor cheerleading. You'll find the practices needed in a world where everyone, eventually, will have to interface with AI in some form. Where ignorance costs people and institutions alike. Where knowing how to work well with these systems isn't optional anymore.

It already isn't.

Reading paths by role

Not every reader needs every chapter on day one. The paths below give you a practical route based on your role. Use them to earn a fast win, then circle back for the rest. Each path still gives you something concrete, but now it is a different kind of outcome: a set of questions tailored to your role that you can use to test your assumptions and sharpen your judgment.

- **Instructor path**, Chapters 7, 8, 5, 9, 12. Redesign one assignment, adopt a verification routine, and publish a clear student policy. Expect a measurable lift in originality and a drop in grading rework.

- **Manager path**, Chapters 1, 3, 4, 13, 14, 15. Set standards, pilot a safe agent, and move the team toward value metrics. Expect better handoffs and fewer fire drills.

- **Creator path**, Chapters 5, 6, 12. Adopt a repeatable writing and visual workflow that protects style and ethics. Expect faster concept cycles without a loss of voice.

- **Student path**, Chapters 7, 5, 8, 9, 13. Learn to defend method, verify claims, and show your thinking. Expect steadier grades and stronger interviews.

- **Researcher path**, Chapters 1, 2, 9, 14, 15, 16. Connect the infrastructure view to cognition, cost, and culture. Expect cleaner methods sections and clearer replication

Late in each chapter you will see a boxed section of questions. Treat them as a pause button. They are there to help you check in with yourself before the next chapter pulls you forward..

A closing word

This book is a snapshot of a moving target. Features shift. Interfaces change. Best practices mature. Many chapters began as articles in my LinkedIn newsletter, Hooked on AI, then were edited into a single arc with a conversational voice and concrete examples. Think curated notebook, not museum catalog. Dip in where the need is most immediate, then circle back for context. If we do this well together, you will leave with a sharper mental model, a sturdier playbook, and a clearer sense of responsibility for the systems you help build and the people they touch.

For more AI content, you can subscribe here, https://www.linkedin.com/build-relation/newsletter-follow?entityUrn=7257851677544710145

1

AI AS A GENERAL PURPOSE TECHNOLOGY

Much like electricity fundamentally reshaped the world, AI is proving to be a general-use technology that is not just changing industries but redefining what it means to be human, to work, and to live in the 21st century. This book is a collection of perspectives gathered from my newsletter, Hooked on AI, which explores AI's impact, and it aims to provide a human-centric view of this unseen revolution. I'll cut through the hype and the fear to explore the practical realities, the ethical considerations, and the profound potential of navigating a world increasingly shaped by intelligent systems. Consider this a guide to understanding our digital entanglement and engaging thoughtfully with the intelligence that is becoming our digital companion.

At some point in the last few years, AI stopped feeling like a clever add-on and started behaving like background radiation. It is no longer just another product on the shelf. It has become part of the air that other products breathe. Economists have a name for technologies that work this way: General Purpose Technologies, or GPTs. They are the tools that turn into conditions. Electricity did it. The steam engine did it. The internet did it. A GPT shows up everywhere, keeps getting better, and becomes more valuable as more of the world plugs into it. When a technology reaches that stage, it is no longer just a feature. It is the environment that everything else is built inside.

You can see the pattern if you walk back through recent history. Electricity began as bright bulbs in a dark room, then crept into factories, then entertainment, then every socket in every wall. The internet started as a nervous little network for researchers, then swallowed commerce, then rewired how people talk, learn, and organize. AI is moving along that same curve, only faster, as if someone leaned on the fast-forward button. It carries information from place to place, nudges decisions, and quietly choreographs workflows we barely notice. It scans images by the millions, scores risk, plans delivery routes, drafts messages, flags odd behavior, and directs human attention to whatever the system finds most urgent. The more we build around it, the more we find ourselves depending on it. The more we depend on it, the more we shape the world to answer to its presence.

Dependence is not a moral failure. It is the standard operating condition of a modern society that runs on invisible pipes. We are already held up by a handful of essential habits that function like socially accepted addictions. Electricity. Clean water. Sewer lines. Air conditioning in cities that would cook without it. The network connections that carry money and medicine and everyday speech. When the power dies, hospitals do not ease into trouble, they snap from tight to fragile in a few minutes. When networks drop, pharmacies cannot check prescriptions, supply chains cannot reconcile inventory, and cashiers can stand there holding real money with no way to complete a sale. We built a civilization that answers to its systems, and people can die when those systems fail. AI is sliding into that layer, joining the list, because it increasingly sits in the loop that keeps those systems moving. The adult question is not how to keep AI out. The adult question is how to run it well enough that the dependence is deserved.

If you want to see where this is headed, follow power and money. Microsoft has signed a long contract with Constellation to bring Three Mile Island Unit 1 back into service, restoring roughly 835 megawatts of nuclear power to feed data centers. That is not the kind of decision anyone makes for a passing fashion. You restart a dormant reactor when you think computational demand is becoming as basic as electricity itself and will remain so for decades. AI does not run on wishes. It runs on energy, and the grid is quietly rearranging itself to keep the new appetite fed.

Capital spending tells the same story in a different language. The largest cloud providers are steering toward record levels of investment tied directly to AI: copilots woven into office software, hosted models, specialized hardware for accelerated compute. Analyst forecasts now place combined AI-related capital expenditures by these platforms in the neighborhood of four hundred billion dollars by the middle of the decade. That is not feature money. That is platform money. When organizations commit at that scale, they are not betting on a gadget. They are betting that AI will underwrite the next era of products, processes, and productivity.

Look sideways across sectors and the pattern resolves into everyday practice. In hospitals, models sort radiology queues, estimate who might deteriorate overnight, and suggest care pathways that clinicians accept or reject with their own judgment. The gain is time and focus, not miracle cures. In logistics, predictive systems help decide where to stage inventory and how to reroute trucks when a storm closes a highway, shaving hours off delays that used to feel inevitable. In classrooms, adaptive practice tools meet students at different levels and hand teachers back hours for feedback, culture, and the kind of encouragement no screen can simulate. In finance, anomaly detectors and scenario models surface risk faster than human analysts, while also tempting institutions to trust whatever glows green on a dashboard a little too much. In energy, grid operators use learning systems to juggle variable wind and solar with the stubborn needs of cities that will not sleep. In public safety, pattern detection and triage tools can speed response, which raises the bar for accountability when they miss. In research and engineering, AI shortens the time between question and prototype, between idea and simulation. It expands the space of possible options. People still decide what is worth doing.

Every new capability brings its own ways to fail. AI does not evaporate human judgment. It accelerates whatever judgment you bring to it. If you are careful, that is good news. If you are careless, bad decisions arrive faster and in greater volume. Biased data can harden yesterday's inequities into tomorrow's default settings. Changes in the real world can drift slowly away from the conditions that a model was trained on, degrading performance in silence. Polished interfaces can hide uncertainty, smoothing the rough edges that would have warned a human user to slow down. As models grow, so do their demands on power and water, often far from the communities that bear the environmental cost. Stack too many mission-critical workflows on a single vendor, and you turn that provider into a single point of failure. The same tools that spot threats can be turned around to probe defenses or poison data. If you cannot trace where inputs came from, you cannot trust the outputs. If you cannot explain how a system failed, you are likely to repeat the same failure when pressure is highest.

All of this is why AI literacy belongs beside reading, writing, and arithmetic, not in some optional elective at the edge. Almost every job will encounter systems that predict, rank, and recommend. People deserve to know what a model was trained to do and what it was not trained to do. They should understand what the system sees clearly, where it is blind, and how uncertain its answers are. They should have a working sense of evaluation metrics, of how performance can drift once a system leaves the lab and starts living in the wild. They should know when human review stays in the loop because the cost of certain errors is lopsided and cruel. They should be able to ask about data lineage, consent pathways, and how someone can appeal a decision. Access is strategy. If only a small group knows how these systems work and how to shape them, that group will set the terms for everyone else.

Governance needs to grow up at the same pace as capability. Rules that scale with risk make sense. High-stakes uses in health, finance, education, and public safety should not be allowed into production on a shrug and a demo. They should face realistic tests before deployment, monitoring and auditing once they are live, and incident reporting so routine and unglamorous that it feels like aircraft maintenance. Procurement contracts should ask about security posture, update practices, supply-chain integrity, and energy disclosures, not just features and price. Clear documentation lowers the barrier for smaller institutions that do not have an army of engineers but still need to evaluate and adopt tools safely. International coordination is not a luxury line item. Models and data cross borders as easily as a rumor. If rules stop at the water's edge, harms will not.

Viewed from abroad, this is clearly not a conversation confined to one country. Saudi Arabia's Vision 2030 treats data and AI like national infrastructure tied directly to state ambitions. The Saudi Data and AI Authority reports that most of Vision 2030's direct and indirect goals lean on data and AI in some form, which turns adoption into a single conversation about policy, education, and build-out, not three separate projects. That strategy is backed by large moves on the ground: multi-billion-dollar investments announced at events such as LEAP, capacity expansions backed by the Public Investment Fund, new AI operators planning data centers at gigawatt scale before the decade is out. When a government binds most of its long-term objectives to data and AI, it is saying that literacy, compute, and governance are not just technical matters. They are civic responsibilities.

People are being prepared in parallel. Saudi Arabia's Ministry of Education has approved a nationwide AI curriculum that is expected to reach more than six million students starting in the 2025–2026 academic year, with SDAIA and partner agencies involved. That is what it looks like when a country treats AI literacy as part of the core civic toolkit rather than an after-hours club. You cannot set the wind in your favor, but you can teach a generation how to handle the sails before the weather changes.

Underneath all this ambition lie physical limits that do not care about strategy decks. Training and serving large models draw real power and drink real water. Efficiency gains are real, yet the total demand will still rise as more sectors plug in. If AI is drifting into the same category as water plants and hospitals, then data centers should be planned and managed with the same seriousness. That means thinking about where they sit in relation to local grids and water tables, being honest about consumption, and designing incentives that reward clean energy and off-peak use. It also means building edge capacity closer to where latency, privacy, or resilience demands local processing. Choices about where we pour concrete and where we pull power will echo for decades in cost and access.

Equity here is not a slogan, it is an engineering requirement. Systems trained on narrow or skewed data will simply not see the communities that are already on the margins. Tools that only wealthy institutions can afford will stretch existing gaps wider. If we want AI to raise the floor rather than raise the ceiling, we need shared public datasets that reflect real populations, evaluation suites that test performance on different groups, and reference implementations that smaller teams can stand up without a heroic budget. Funding has to reach clinics, schools, and local governments that carry heavy workloads with thin resources. AI will not, on its own, equalize outcomes. What it can do is make expert time and attention go further, which is a powerful lever if we choose to pull it for the places that need it most.

Work will change shape. That part is not in serious doubt. Routine cognitive tasks, the predictable legwork of many jobs, will migrate toward machines. In their place will come roles that focus on system design, curation, safety, evaluation, incident response, and the very human art of teaming with AI instead of being replaced by it. That transition should not be treated as a private hobby. It deserves the same level of planning we bring to roads or power lines, with reskilling that is paid, practical, and tied to real jobs rather than vague promises. A society that treats people as the flexible, disposable part of the system will find itself fragmented. A society that treats learning as shared infrastructure will see returns that compound.

Resilience means planning for the dead spots, not just the uptime graphs. If AI sits inside a workflow that matters, the manual fallback should exist before day one, not after the first outage. Failover should be practiced with the same unglamorous repetition as a fire drill. Records should be kept in formats that remain usable if a vendor disappears or a license changes. Decision logs should be human-readable without needing a specialist on call at three in the morning. Preparedness is not a slogan. It is the quiet routine that keeps working when the lights flicker.

Taken together, these threads point to a simple imperative. Learn the vocabulary. Learn the limits. Learn the levers. The future will not be interested in whether we used AI. It will be interested in whether we used it wisely. A simple image helps. You do not control the wind. You control your sails. The wind in this case is the accelerating capability of AI. The sails are education, design choices, institutional practice, and public policy. Adjust them early. Adjust them often. Aim for human outcomes that are better because machines helped, not systems optimized for machines with humans trotting behind.

There is also a cultural shift here that deserves to be named plainly. We used to treat AI as an experiment at the edge of the system, something interesting in the lab and entertaining in demos. It is now moving into the company of electricity, water, and networks, part of the critical stack that keeps doors open and services running. That status calls for humility, transparency, and care. As McLuhan warned, we shape our tools, then our tools shape us. AI is a tool that answers back. It alters pace, patterns, and power. If we integrate it with equity, accountability, and an honest reckoning with energy and infrastructure, we widen opportunity and smooth some of the brittleness in our systems. If we let it concentrate advantage without guardrails, we will harden fragility and shrink the circle of people who benefit.

Falling behind here does not mean missing a trend, it means losing the ability to steer. This is not a call to fall in love with new code. It is a call to fluency. Learn enough to pair human strengths with machine capabilities. Ask for documentation. Ask what happens when things break. Ask for ways to opt out in settings where harm cuts deep. Build on open standards where you can, and demand real export paths when you cannot. If you cannot move your data or your models, you are not steering your own future. The individuals, institutions, and societies that treat AI literacy as essential, not optional, will keep their footing while the ground moves beneath them.

AI is entering the same club as electricity and the internet because it is becoming a precondition for how work gets done and how services reach people. That is a high honor and a hard responsibility. The test is straightforward. Does a deployment make outcomes better, fairer, and more resilient, or does it only make them faster, shinier, and more fragile. Choose as if outage were a matter of "when," not "if," because that is the reality of all systems. Build as if the next user is your most vulnerable neighbor. Do that, and the dependency we are weaving into our infrastructure will be one we can defend without flinching.

Hands on the Wheel

AI is becoming part of the background, like power or plumbing. Before you hand the wheel to anything, it helps to be honest about where you already rely on it and where you refuse to. Use these questions as a quiet audit, not a contract.

1. Where in my day is AI or invisible automation already making decisions for me, even if I rarely notice it?

2. If I treat AI as infrastructure instead of a gadget, what responsibilities does that put on me as a designer, teacher, or manager?

3. Which kinds of decisions in my work should never be delegated to a system, no matter how accurate it becomes?

4. When I imagine "keeping a human in the loop," what does that actually look like in my own context, not just as a slogan?

2

The Quiet Revolution

If AI is indeed the new electricity, then its initial "appliances" are already appearing in our daily work lives. Following the broad declaration of AI as a foundational technology, this chapter zooms in on a tangible, immediate impact: the quiet revolution happening in the digital workplace. We'll explore how AI is fundamentally altering how we interact with software, demanding and defining a new kind of digital proficiency. It's a shift that shows AI isn't just a theoretical concept, but a present reality already changing practical skills and reshaping the landscape of our professional lives.

In the spring of 2023, a graphic designer named Alex sat in her office in downtown Chicago staring at a blank screen. She had been at the same marketing agency for six years. She knew Photoshop the way a pianist knows a keyboard, every shortcut memorized, every tool mastered through thousands of hours of practice. She was, by any traditional measure, an expert. As I think back to my own early encounters with these tools, I remember the quiet thrill of the blank screen and the sense that something both familiar and startling was about to happen.

Her assignment that morning was straightforward: create a logo concept for a healthcare startup. Blue gradients. Soft edges. Modern feel. The kind of work she had done hundreds of times before.

But something strange happened. Her junior colleague, a recent graduate named Marcus who had been at the agency for all of four months, finished first. Not just first, either. He produced twelve polished variations in the time it took Alex to complete two.

Marcus was not more talented than Alex. He did not work harder. He simply typed a sentence into a text box: "modern healthcare logo with blue gradients and soft edges." Adobe's new AI system, Firefly, did the rest.

Here is the question that matters: Was Marcus cheating? Was he being lazy? Or was he doing something that Alex, for all her expertise, had not yet learned how to do?

The answer, it turns out, tells us something important about what it means to be good at your job in an age when the software we use has quietly started thinking for itself.

To understand what happened to Alex, we need to step back and consider a word that gets thrown around constantly in professional settings: proficiency.

For decades, proficiency in software meant something specific. It meant knowing where to find things. It meant memorizing keyboard shortcuts. It meant understanding, at a deep level, how a program worked so you could bend it to your will. The proficient professional was the one who had put in the hours, who had developed muscle memory, who could navigate complex menus without thinking.

This kind of proficiency was hard-won. It created clear hierarchies. The person who knew Excel's advanced formula syntax had an advantage over the person who did not. The designer who could execute a complex selection in Photoshop by hand was more valuable than one who could not. Experience, time, and sustained practice all played a role in that value.

But here is the strange part. Over the past two years, the major software platforms that define professional work have undergone a quiet revolution. Microsoft embedded an AI system called Copilot into Word, Excel, and PowerPoint. Adobe wove its Firefly engine throughout Photoshop and Illustrator. Google integrated Gemini into Gmail, Docs, and Sheets. These were not separate applications requiring new expertise. They were features tucked inside familiar programs, waiting to be discovered.

And they changed everything.

Consider what Copilot can do inside a Microsoft Word document. It does not just check your spelling. It drafts content based on a few sentences of instruction. It summarizes a forty-page report into three paragraphs. It suggests whether your tone is appropriate for your audience. It learns your organization's terminology and maintains consistency across everything you write.

Or think about what happens when a financial analyst opens Excel now. Copilot can look at a spreadsheet full of numbers and tell you what patterns exist. It recommends formulas. It generates charts that communicate the story hidden in the data. The analyst who once spent hours manipulating cells can now spend that time thinking about what the numbers actually mean.

This is more than a minor upgrade. It represents a fundamental shift in what software does. For fifty years, software functioned as a tool. You told it what to do, step by step, and it executed your commands. The new paradigm is different. Software has become a collaborator. You tell it what you want, and it figures out how to get there.

The implications of this shift are profound, and they are not what most people expect.

You might assume that AI integration makes everyone's job easier, levels the playing field, and reduces the importance of expertise when anyone can type a prompt and get a result.

The research tells a more complicated story.

In 2024, the Bureau of Labor Statistics reported a 2.3 percent productivity increase in the nonfarm business sector. That sounds like good news. AI tools are making people more productive. Except when researchers looked more closely at how workers were actually using these tools, they found something puzzling.

Early adopters were not immediately more productive. They worked longer hours. They reported decreased focus time. They struggled. The problem was not that the tools did not work. The problem was that using them effectively required a skill that nobody had taught these workers how to develop.

Think about what it takes to get good results from an AI system embedded in your software. You need to know what to ask for and phrase your request in a way the system understands. You need to evaluate the output critically, catching errors and biases that the AI cannot see. You need to know when to accept a suggestion and when to reject it.

This is a new kind of expertise. Call it prompt fluency, or AI collaboration, or human-machine partnership. Whatever the label, it represents a capability that did not exist five years ago and that most training programs still do not teach.

The research found something else, too. High-performing teams were twice as likely to use AI tools effectively as average teams. The difference was not access to the technology. Everyone had access. The difference was in how people approached the partnership.

Let me tell you about Daniel, a financial analyst at a mid-sized investment firm in Boston. Before Copilot, Daniel's job followed a predictable rhythm. He would receive data. He would spend hours cleaning it, organizing it, running calculations, building charts. By the time he finished the mechanical work, he often had little energy left for the part of his job that actually mattered: interpreting what the numbers meant and advising clients on what to do.

Daniel started experimenting with Copilot in late 2023. At first, he was skeptical. He had spent years developing his Excel skills, and he worried that relying on AI would make him lazy. He worried his skills would atrophy.

But something unexpected happened. As Copilot took over the mechanical tasks, Daniel found himself with time to think. He started noticing patterns in the data that he had missed before, not because the patterns were hidden, but because he had always been too busy with the mechanics to look for them. His reports became more insightful. His recommendations became sharper. He started delivering work ahead of schedule, which meant clients had more time to act on his advice.

Daniel did not become less skilled. He became differently skilled. The value he provided shifted from data manipulation to data interpretation, from execution to strategy.

This is the counterintuitive insight at the heart of the AI integration story. The tools do not replace human expertise. They redirect it. They free professionals from mechanical tasks so they can focus on judgment, creativity, and the kinds of complex thinking that machines still cannot do.

But that redirection only works if you know how to manage the partnership.

To appreciate how far we have come, consider a piece of software history that many professionals would rather forget.

In 1997, Microsoft introduced an animated paperclip named Clippy into its Office suite. Clippy was supposed to help. It would pop up when you started writing a letter and ask, "It looks like you're writing a letter. Would you like help?" The problem was that Clippy almost never understood what you actually needed. Its suggestions were generic, interruptive, and frequently wrong. Users hated it. Microsoft eventually removed it.

Clippy failed because it operated from a limited database of pre-programmed responses. It could recognize certain patterns, but it had no understanding of context, intent, or meaning. The disconnect between what users needed and what Clippy could offer made the assistant worse than useless.

The distance between Clippy and Copilot is the distance between a parlor trick and genuine intelligence.

Modern AI assistants are built on large language models trained on vast amounts of text. They do not just recognize patterns. They understand context. They learn from interactions. They adapt to individual work styles. They can interpret ambiguous requests and generate novel solutions to problems they have never seen before.

This technological leap explains why the current moment feels different from previous waves of automation. We are not talking about software that executes commands faster. We are talking about software that understands what you are trying to accomplish and helps you get there.

The risks are real, and they matter, so we need to address them.

The most immediate concern is quality control. AI systems are impressive, but they make mistakes. They generate content that sounds authoritative but contains factual errors. They produce work that reflects biases present in their training data. They sometimes miss nuances that a human expert would catch immediately.

This means that the time saved through AI assistance must be partially reinvested in verification. You cannot simply accept what the machine produces. You have to check it, especially in high-stakes contexts like legal documents, medical communications, or financial advice. Professionals who skip this step will eventually get burned.

There is also the dependency problem. As professionals become accustomed to AI handling certain tasks, their fundamental skills in those areas may weaken. A designer who always uses generative fill might lose the ability to make complex selections by hand. An analyst who relies on Copilot for formulas might forget how the underlying math works.

This creates vulnerability. What happens when the AI tools are unavailable? What happens when they produce obviously wrong results and you lack the expertise to recognize the error? Maintaining core competencies while leveraging AI enhancement requires deliberate effort and planning.

And then there are the organizational questions. Who owns AI-generated content? What happens to sensitive data processed by these systems? How do you ensure equitable access to AI-enhanced tools across different levels of an organization? These are not simple questions, and most companies are still figuring out the answers.

Let me return to Alex, the graphic designer in Chicago, because her story has a second act.

After watching Marcus produce those twelve logo variations, Alex did not dismiss what she saw. She did not retreat into resentment about how things used to be. Instead, she got curious.

She started experimenting with Firefly. At first, her results were mediocre. She would type a prompt and get something generic, something that looked like it could have been made by anyone. But over time, she learned something important. The quality of the output depended entirely on the quality of her input.

Her six years of design expertise had not become irrelevant. They had become essential in a new way. She knew what made a logo work. She understood color theory, visual hierarchy, the subtle differences between fonts that convey trust versus fonts that convey innovation. She could look at an AI-generated design and immediately see what was wrong with it, what needed adjustment, what direction to push the next prompt.

Marcus could generate twelve variations. But Alex could generate twelve variations and then refine the best one into something that actually solved the client's problem. She could explain why certain design choices worked and others did not. She could use the AI as a tool to explore possibilities faster, while bringing human judgment to the final decisions.

Within six months, Alex was producing work that was both faster and better than before. The AI had not replaced her expertise. It had amplified it.

Here is the pattern we are seeing across industries.

Lawyers use AI to distill lengthy documents into actionable insights, then apply their legal judgment to determine what matters. Architects generate conceptual visualizations to explore possibilities quickly, then bring their spatial reasoning to refine the designs. Healthcare administrators draft patient communications with AI assistance, then add the empathy and context that machines cannot provide.

In every case, the professionals who thrive are the ones who understand what they bring to the partnership. The AI handles speed, scale, and pattern recognition. The human provides judgment, creativity, ethics, and the ability to understand what another human actually needs.

This division of labor is not static. As AI systems become more sophisticated, the boundary will shift. Tasks that require human judgment today may be handled by machines tomorrow. But the fundamental insight will remain: the value of human professionals lies not in executing mechanical tasks but in providing the things that machines cannot.

So what should you do if you are a professional navigating this transition?

The first thing to understand is that the AI features are already in your software. You do not need to buy new tools or learn new platforms. Start by exploring what is already there. Ask Copilot to summarize a document. Have Gemini draft an email response. Use Firefly to generate an initial design concept. Low-stakes experimentation builds comfort and understanding.

The second thing is to practice the skill that matters most: communicating clearly with these systems. When you ask an AI to produce something, be specific. Tell it the tone you want, the format you need, the audience you are writing for. The more precisely you can describe what you want, the better the results will be.

The third thing is to maintain your critical faculties. Do not accept AI output uncritically. Check facts. Evaluate suggestions against your expertise. Remember that you are the one with judgment and context. The AI is a tool, a remarkably powerful tool, but still a tool.

And finally, keep developing your fundamental skills. The professionals who will be most valuable in an AI-augmented world are not the ones who know how to use AI. Everyone will know how to use AI. The valuable professionals will be the ones who bring genuine expertise to the partnership, who can evaluate AI output against deep knowledge, who can see what the machine misses.

We are at the beginning of something significant. The integration of artificial intelligence into the software we use every day is not a trend that will pass. It is a fundamental shift in how professional work gets done.

The question facing every professional now is not whether AI will become part of their daily experience. That question has already been answered. The question is how they will adapt to a world where proficiency means something different than it used to.

For fifty years, we learned to use software by memorizing commands, practicing techniques, and developing muscle memory. The new skill is harder to define. It involves communication, judgment, partnership, and a willingness to let go of mechanical tasks so you can focus on the work that actually matters.

Alex learned this lesson in a Chicago office, watching a junior colleague outpace her with a single sentence typed into a text box. Daniel learned it in Boston, discovering that the AI freed him to do the part of his job he had always wanted to do. Professionals across every industry are learning it now, some faster than others, some more willingly than others.

The transformation is quiet. It arrives not with fanfare but with subtle implementation, a new button here, a new feature there, until one day you realize that the software you have been using for years has become something else entirely.

It has become a partner. Learning to work with that partner may be the most important professional skill of the next decade.

Copilot, Quiet Questions

Embedded AI in tools like Word, Excel, and Photoshop is quietly redefining what it means to be 'good with software.' These questions invite you to examine what kind of skill you want to have in a world of copilots.

1. When I say someone is "proficient" with a tool now, what do I actually mean: they know the menus, or they know how to ask good questions of the copilot?

2. Which of my current skills would I be afraid to lose if a copilot did most of the keystrokes for me?

3. Is there anywhere in my work where I am hiding behind "the tool did it" instead of owning the judgment behind the result?

4. If my students or colleagues grow up never touching the "manual" features, what do I think they will miss, and what might they actually gain?

3

The Age of AI Agents

Building on the idea of changing software proficiency, this chapter looks at the larger implications for work and the economy. The shift in individual skills is a microcosm of a much larger transformation driven by AI agents. These agents are not just changing how we work, but what work looks like across industries and the global economy. As we delve into the age of AI agents, we'll address anxieties about job displacement and explore how this technology is redefining work, industry, and the global economy, prompting a reorientation rather than just a reduction in roles.

The Agentic Era: Work Reimagined

What does it mean to work when the nature of work itself is being rewritten?

We are living through a transformation more profound than the shift from the industrial age to the digital era. This isn't merely technological evolution. It's a fundamental reordering of human purpose, structure, and scale. Artificial Intelligence, particularly agentic systems capable of acting autonomously, is not just I often think back to conversations with students who sense both excitement and unease about these shifts, a reminder that progress is as much about feeling as it is about technology.accelerating productivity; it is reshaping how we define work, what it means to be employed, and how value is created in the 21st century.

The implications extend beyond job descriptions and organizational charts. We're witnessing the emergence of entirely new forms of professional identity, where human capability becomes inseparable from the AI systems we train, refine, and deploy. The worker of tomorrow may not be hired as an individual, but as an integrated human-AI partnership that has been cultivated over years of collaboration.

From Linear to Exponential

"We shape our tools, and thereafter our tools shape us."

Marshall McLuhan

For centuries, productivity scaled linearly: more people, more hours, more output. This was the logic of the industrial age, carried forward into the digital era. AI agents invert that fundamental assumption. Tasks that once required teams of ten, or even a hundred, can now be orchestrated by a handful of workers, each managing dozens of AI agents operating in parallel across multiple workflows.

The numbers tell a striking story. According to the World Economic Forum, 83 million jobs are expected to be displaced, but 69 million new roles will be created by 2027. At the same time, machines currently perform 34% of all tasks, a figure expected to rise to 42% by 2027. This isn't a story of obsolescence, but of reorientation.

What emerges is not a world with less work, but a world where work itself is fundamentally reconceptualized. The unit of productivity changes. Instead of measuring output per person-hour, we begin measuring output per orchestrated system. The human role shifts from executor to architect, from typing every line of code to designing the systems that generate, validate, and refine entire workflows.

This transformation is already visible in early adopter organizations. Software development teams are transitioning from writing code line-by-line to architecting systems that generate, test, and deploy code automatically. Marketing professionals are evolving from content creators to content orchestrators, managing AI systems that produce dozens of campaign variations simultaneously. Financial analysts are becoming model architects, designing AI systems that process market data and generate insights at superhuman scale.

But what does this mean for human agency? Are we becoming conductors of an increasingly automated orchestra, or are we being conducted by it?

The Paradox of Human Relevance

> "The real problem is not whether machines think, but whether men do."

B.F. Skinner

In this new landscape, the most valuable human skill isn't technical proficiency, it is systems thinking. The ability to break complex problems into modular tasks that can be delegated to autonomous agents, validated for accuracy, and woven back into coherent solutions. Coding won't disappear, but it will become more about logic and design than syntax.

This represents a profound cognitive shift. We're not just learning to use new tools; we're learning to think in new ways. The essential literacy of the future may not be traditional programming, but rather the ability to understand abstraction, orchestration, and verification at scale.

The most successful professionals are already developing what we might call AI fluency: the capacity to understand not just what AI can do, but when and how to deploy it most effectively. This includes recognizing the strengths and limitations of different AI systems, crafting effective prompts and instructions, and maintaining quality control over automated processes.

> "Learning to code may no longer be about syntax. It's about learning how to think."
>
> A synthesis of Peter Drucker and Alan Kay

Economic Expansion, Not Contraction

"Automation applied to an efficient operation will magnify the efficiency; applied to an inefficient operation it will magnify the inefficiency."

Bill Gates

The focus on job displacement misses a larger economic truth: AI doesn't just substitute for labor, it expands what's economically possible. According to PwC, AI could contribute $15.7 trillion to global GDP by 2030, making it the single biggest commercial opportunity of our era. McKinsey's research suggests generative AI alone could deliver $2.6 to $4.4 trillion in value annually across industries from finance to healthcare to manufacturing.

Yet these numbers, impressive as they are, don't capture the deeper transformation. What happens when the cost of intellectual labor approaches zero? When internal tools for a company can be built in days rather than quarters? When a one-person startup can wield the productive output of what once required a hundred-person team?

The answer isn't economic contraction. It's economic expansion. Entire categories of previously unfeasible work become viable. Personalized education, customized healthcare, hyperlocal services, and tailored enterprise solutions all shift from impossible to inevitable.

Consider the emergence of micro-enterprises that can compete with traditional corporations despite having only a handful of human employees. These organizations leverage AI agents to handle customer service, content creation, data analysis, and even product development, allowing human founders to focus on strategy, relationships, and creative vision.

"The best way to predict the future is to invent it."

Alan Kay

The Dual Nature of Organizational Evolution

Perhaps nowhere is this transformation more visible than in the changing structure of organizations themselves. We're witnessing the simultaneous rise of the microenterprise and the superpowered corporation. The barrier to entry for launching a startup is falling dramatically. One person with a laptop and an army of AI agents can now achieve what once required entire departments.

Yet large enterprises aren't shrinking, they're scaling differently. Through parallel task delegation, 10,000 engineers can become 100,000 digital laborers. Despite fears of mass layoffs, enterprise surveys show that 78% of businesses now use AI in at least one function, with over $300 billion in global AI spending projected by 2026. Most are choosing capability expansion over cost-cutting.

This divergence creates interesting dynamics in the job market. While some roles are being automated away, new categories of work are emerging: AI trainers, prompt engineers, automation architects, and human-AI collaboration specialists. These roles require deep understanding of both human psychology and AI capabilities.

What does this say about the future of human collaboration? Are we entering an age of enhanced collective intelligence, or increasingly isolated individual productivity?

The Question of Accountability

"The future is already here, it's just not evenly distributed."

William Gibson

This brings us to one of the most challenging aspects of the agentic era: accountability. AI agents make decisions, but when they make mistakes, who bears responsibility? In high-stakes domains like law, medicine, and finance, the chain of responsibility cannot be allowed to become opaque.

Until these systems achieve reliability that approaches human-level consistency, humans remain accountable for the outputs. It's not the AI's job to be productive. It's your job to make the AI productive.

This creates a new kind of professional responsibility, not just for our own actions, but for the actions of the systems we deploy and oversee. It's a form of accountability that requires both technical understanding and ethical judgment.

The legal and regulatory frameworks are still catching up to this reality. Professional licensing bodies are grappling with questions about liability when AI systems make errors. Insurance companies are developing new products to cover AI-related risks. Organizations are implementing new governance structures to ensure human oversight of automated decision-making.

Memory, Identity, and the Cognitive Passport

With the rise of persistent memory in AI agents, we're entering uncharted territory around identity and ownership. Imagine a work agent that remembers every line of code you've written, your communication style, your decision-making patterns across years of collaboration. This isn't just a tool, it is a digital extension of professional identity.

But what happens when you change employers? Who owns that accumulated knowledge? Should your next company inherit your digital cognitive patterns? We may need to develop what could be called cognitive passports: standardized, portable memory containers that allow workers to transfer trusted context across platforms and domains.

These cognitive passports would represent a fundamental shift in how we think about professional credentials. Instead of static résumés that list past experiences, we would have dynamic, interactive representations of our working relationships with AI systems. A cognitive passport might include collaboration history showing records of successful projects completed with AI assistance, communication patterns demonstrating the ability to work effectively with different types of AI systems, quality metrics tracking output accuracy and validation processes, specialization areas documenting expertise in training AI for specific domains, and ethical frameworks providing evidence of responsible AI governance and oversight.

In a world where our AI agents know us better than we know ourselves, what does professional identity even mean?

The AI Agent as Professional Asset

Perhaps the most revolutionary aspect of this transformation is emerging in the hiring process itself. We're approaching a future where employers won't just be evaluating your skills, experience, and cultural fit, but also the capabilities and quality of the AI agents you've trained to work alongside you.

Consider this scenario: Sarah, a marketing professional, has spent three years training her AI agent to understand her company's brand voice, customer segments, and campaign strategies. The agent knows her creative process, can anticipate her needs, and has learned to produce initial drafts that require minimal revision. When Sarah applies for a new position, she's not just offering her individual expertise. She's offering the productivity multiplier of a finely-tuned human-AI partnership.

Future job applications might include an AI Agent Portfolio alongside traditional résumés. This portfolio would demonstrate agent capabilities showing what specific tasks and workflows the AI can handle independently, training quality with metrics showing the accuracy and reliability of the AI's outputs, collaboration efficiency with data on how effectively the human-AI team works together, domain expertise providing evidence of the AI's specialized knowledge in relevant fields, and adaptability showing the agent's ability to learn new processes and integrate with different systems.

This creates a new employment model where companies hire human-AI packages rather than just individuals. The value proposition becomes exponentially higher. Instead of hiring someone who can produce X amount of work, companies can hire someone whose AI partnership can produce five or ten times the work while maintaining quality and strategic oversight.

This shift fundamentally changes salary negotiations and job market dynamics. A professional with a highly trained, specialized AI agent becomes significantly more valuable than someone with equivalent human skills but no AI partnership. The agent itself becomes a form of intellectual property and competitive advantage.

Organizations will need to develop new processes for integrating not just new human employees, but their AI agents as well. This might include security clearance to ensure AI agents meet company data protection standards, system integration connecting personal AI agents with company infrastructure, knowledge transfer allowing AI agents to learn company-specific processes and information, and performance benchmarking establishing metrics for human-AI team productivity.

The Invisible Infrastructure

"What the computer is to me is the most remarkable tool that we have ever come up with. It's the equivalent of a bicycle for our minds."

Steve Jobs

We often think of AI as a tool layered on top of existing systems. But increasingly, it resembles something more fundamental: a new cognitive infrastructure that underlies the entire economy. Just as electricity or the internet transformed every industry by becoming invisible and ambient, AI agents are becoming the substrate on which all other work depends.

"Civilization advances by extending the number of important operations which we can perform without thinking about them."

Alfred North Whitehead

This invisibility is perhaps the most profound aspect of the transformation. We're not just automating tasks. We're redefining what constitutes a task. AI agents change the fundamental unit of productivity, reshape workflows, compress timelines, and enable organizations to solve problems that were once considered too small or too complex to tackle.

The infrastructure metaphor is particularly apt because, like electricity or internet connectivity, AI capability is becoming a baseline requirement rather than a competitive advantage. Organizations that fail to integrate AI effectively will find themselves at a fundamental disadvantage, much like businesses that tried to operate without electricity or internet access in previous technological transitions.

The Skills Arms Race

As AI agents become more central to professional value, a new skills economy is emerging around human-AI collaboration. The professionals who thrive will be those who can develop AI training expertise, understanding how to effectively teach AI systems domain-specific knowledge, company procedures, and quality standards. They will also learn to master prompt engineering, crafting precise, effective instructions that guide AI systems toward desired outcomes, combining technical knowledge with communication skills and deep understanding of AI capabilities. They will build quality assurance systems, developing processes to validate AI outputs, catch errors, and ensure consistent quality, drawing on domain expertise and systems thinking. They will learn to navigate ethical boundaries, understanding when and how to use AI appropriately, maintaining human oversight in critical decisions, and ensuring responsible deployment of automated systems. Finally, they will orchestrate complex workflows, managing multiple AI agents working in parallel, coordinating handoffs between automated and human tasks, and maintaining overall project coherence.

These meta-skills for managing AI relationships are becoming as valuable as traditional domain expertise. In many cases, they're becoming more valuable because they multiply the impact of domain knowledge exponentially.

The Human Frontier

"Every great advance in science has issued from a new audacity of imagination."

John Dewey

In healthcare, AI reduces administrative burden and shortens wait times. In education, it enables personalized learning at unprecedented scale. In research, it simulates experiments, synthesizes literature, and generates hypotheses. For the first time in history, we're seeing venture-backed startups with fewer than ten employees raise millions in funding because AI agents can accomplish what once required entire floors of developers.

The democratization effect is particularly striking in creative fields. Independent filmmakers can now produce visual effects that rival studio productions. Solo musicians can create orchestral compositions. Individual researchers can analyze datasets that once required entire academic departments.

But perhaps the most significant change is not in what we can do, but in what we choose to do. The real revolution is not in the technology itself, but in how we decide to use it.

The Competitive Landscape

The bifurcation between individual AI agents and enterprise AI systems is creating interesting competitive dynamics. While large organizations have access to more resources and data, individuals with well-trained personal AI agents can compete in ways that were previously impossible.

For individuals, this creates opportunities including lower barriers to starting businesses, the ability to compete with larger organizations in specific niches, higher productivity and earning potential, and more flexibility and autonomy in work arrangements. For organizations, this creates challenges including retaining employees who become more valuable as independent operators, competing with individuals who have lower overhead costs, managing intellectual property when employees own their AI agents, and adapting hiring and compensation models for human-AI packages.

The Choice Before Us

> "The significant problems we face cannot be solved at the same level of thinking we were at when we created them."
>
> Albert Einstein

AI is not the end of work. It's the beginning of a new relationship between human creativity and computational power. The greatest contribution we can make going forward is not to compete with machines, but to imagine more meaningful work, design better systems, and solve problems we once considered beyond our reach.

The question isn't whether AI will change how we work. It's whether we'll have the wisdom to shape that change in ways that amplify human flourishing rather than diminish it.

This wisdom must extend to how we handle the transition period. As AI agents become integral to professional identity, we need to ensure that access to AI training and partnership opportunities doesn't create new forms of inequality. The benefits of human-AI collaboration should be distributed broadly rather than concentrated among those with existing advantages.

Governments and institutions will need to address education reform by updating curricula to include AI collaboration skills, professional standards by developing certification programs for human-AI partnerships, economic policy to ensure the benefits of AI productivity gains are shared equitably, labor rights to protect workers' rights to their AI agent relationships and cognitive passports, and antitrust concerns to prevent concentration of AI capabilities among a few large platforms.

The Path Forward

The future of work isn't about what machines can do. It's about what we choose to become. The most successful individuals and organizations will be those who embrace the concept of augmented humanity, where human creativity, judgment, and values guide increasingly powerful AI capabilities.

This future requires new forms of literacy, new professional identities, and new social contracts. But it also offers unprecedented opportunities for human flourishing. When routine cognitive tasks are automated, human energy can be redirected toward creativity, relationship-building, ethical reasoning, and solving complex problems that require wisdom rather than just intelligence.

The professionals who thrive in this new landscape will be those who see AI not as a threat to human relevance, but as an amplifier of human potential. They will build AI agents that extend their capabilities while remaining grounded in human values and judgment.

The agentic era is not about replacing humans with machines. It's about creating new forms of human-machine partnership that allow both to contribute their unique strengths to solving problems and creating value. The future belongs not to humans or AI, but to the intelligent collaboration between them.

As we stand at the threshold of this transformation, the question isn't whether to embrace AI agents as professional partners, but how to do so in ways that enhance rather than diminish human agency, creativity, and dignity. The choices we make today about training, deploying, and governing AI agents will shape the nature of work and human purpose for generations to come.

Agentic Era Questions

AI agents promise to watch inboxes, close tickets, and move work through systems on our behalf. The temptation is to either panic or outsource everything. Sit with where you would actually be comfortable letting an agent act for you.

1. Which parts of my job are truly about judgment and which are mostly coordination and follow-through?

2. If an AI agent could reliably handle those coordination tasks, what would I say my real job is?

3. Where would I feel uneasy letting an agent act in my name, even if it saved time? What does that discomfort reveal about my values?

4. How would I want accountability to work when an agent makes a mistake that affects other people?

4

THE GREAT PROMOTION

This chapter offers a more hopeful perspective on the future of work discussed previously. While the rise of AI agents may raise anxieties about job displacement, this article presents a counter-narrative. It suggests that instead of a reduction in roles, we might be on the cusp of a "great promotion," where AI elevates human potential and shifts our collective role towards management and higher-level tasks. Following the discussion on the changing nature of work, this chapter explores how AI could transform everyone into managers, offering a more optimistic and empowering perspective on the future of employment.

Remember when farmers first got their hands on tractors? Suddenly, one person could do the work of many, and something interesting happened: people had time to explore new possibilities. Some of those farmers' kids became mechanics, others started businesses in town, and a few might have even pioneered the first barbecue cookoffs. That same pattern of technological advancement creating new opportunities is about to repeat itself with artificial intelligence, only this time at a far greater scale.

As a teacher, I've watched students light up when they grasp that shift—not a surrender to machines but a chance to redefine what meaningful work looks like. There's a common fear that AI will replace human workers, leaving masses unemployed. But what if we're looking at it all wrong? What if, instead of replacing us, AI is about to give nearly everyone a promotion?

The New Workplace Hierarchy

Think of today's workplace like a pyramid. At the base, you have people performing the foundational work, including processing data, handling routine tasks, and executing day-to-day operations. Above them are managers who oversee these operations, ensure quality, and make strategic decisions. What's about to happen is that AI will slide in at the bottom of this pyramid, and everyone else will move up a level.

This isn't science fiction; it has already begun. Instead of replacing human workers, AI is becoming a new type of employee that needs human oversight. Just as a manager today might oversee a team of ten people, tomorrow's workers will manage a team of AI agents, making sure they're doing their jobs correctly, efficiently, and ethically.

Imagine you're someone who spends most of your day working on a laptop, processing documents, analyzing data, or handling customer inquiries. In the AI-enhanced future, instead of doing all these tasks yourself, you'll be overseeing a small team of AI agents that handle the basic work. Your job shifts from being the doer to being the manager who ensures everything is running smoothly.

This raises a natural question: can you really have that many managers? Counterintuitive as it sounds, yes. As AI handles more basic tasks, it creates the capacity for exponential growth in what we can accomplish. When one person can oversee multiple AI agents doing the work that used to require several humans, we can tackle more projects, explore more ideas, and build more businesses than ever before.

Despite AI's capabilities, certain professions will always need human hands and minds. Plumbers, electricians, and other skilled trades aren't going anywhere. You can't fix a leaky pipe with an algorithm. Even in these fields, though, AI will serve as a powerful assistant, helping with scheduling, ordering parts, diagnosing problems, and handling paperwork, allowing these professionals to focus more on their specialized skills. They become managers of their AI tools rather than being replaced by them.

> "Our intelligence is what makes us human, and AI is an extension of that quality."
>
> Yann LeCun

New Roles, New Possibilities

As AI systems become more prevalent, entirely new categories of jobs will emerge. Data Quality Managers will ensure AI systems are learning from accurate, up-to-date information, teaching AI about new developments and helping to humanize the data it processes. AI Oversight Specialists will function like air traffic controllers, monitoring AI operations and stepping in when something needs adjustment. Prompt Engineers will specialize in communicating with AI systems, translating human intentions into effective instructions. Ethics Auditors will monitor AI decisions and outputs, checking for bias, fairness, and alignment with human values.

Perhaps the most exciting aspect of this transformation is how it will help bring more ideas to life. Today, most of us have countless ideas that never see the light of day simply because we don't have the time or resources to pursue them. AI changes that equation.

Imagine you have an idea for a movie. Right now, turning that idea into reality would require enormous resources, connections, and time. In the future, AI could help you draft the script, identify potential investors, suggest casting options, and handle much of the logistical planning. Your role would be to guide this process, make key creative decisions, and ensure the final product matches your vision. The same pattern applies across industries. Whether you're starting a business, organizing an event, or developing a new product, AI can handle much of the groundwork so you can focus on the higher-level decisions that require human judgment and creativity.

For those who are already experts in their fields, AI becomes a force multiplier. A brilliant director might only be able to make one movie every few years today because of the sheer amount of work involved. With AI assistance, they could potentially triple or quadruple their output while maintaining their creative vision and quality standards. Scientists could run more experiments, entrepreneurs could launch more businesses, and educators could reach more students. In all of this, humans remain in control, directing these AI tools toward meaningful goals while ensuring the quality and integrity of the output.

Managing the Transition

This shift won't happen overnight, and it won't always be smooth. Organizations will need to invest in training programs to help workers develop the skills needed to manage AI systems effectively. That includes understanding AI capabilities and limitations, learning to give clear instructions to AI systems, developing oversight and quality control processes, making ethical decisions about AI use, and maintaining human connections in an AI-enhanced workplace.

As AI takes on more basic tasks, we're likely to see significant economic growth. When routine work is handled more efficiently, it creates space for new industries and opportunities we haven't even imagined yet. Just as the automation of farming led to the growth of new industries like entertainment and finance, AI automation will open room for new economic sectors.

It's natural to feel some anxiety about these changes. The future can be unsettling, especially when it involves something as powerful as AI. Much of this fear, though, stems from uncertainty rather than actual threats. As we become more familiar with AI's capabilities and limitations, we'll see it more clearly for what it is: a powerful tool that, when properly managed, can enhance rather than replace human capabilities.

The future of work isn't about humans versus AI. It's about humans and AI working together, with each focusing on what they do best. AI will handle the routine, repetitive tasks, while humans move up to focus on oversight, creativity, and strategic thinking. This is not just a change in how we work; it is, in a very real sense, a promotion for humanity as a whole.

As we navigate this transition, the priority must remain human agency and control. AI should enhance our abilities, not replace our judgment. By embracing this new role as managers and guides of

AI systems, we can create a future where technology serves human flourishing rather than the other way around.

The greatest challenge, and opportunity, ahead of us isn't preventing AI from taking our jobs. It is preparing ourselves to step into these new, more elevated roles. The future of work is calling, and it's offering nearly everyone a promotion.

Stewardship Check-In

As more tasks become assistable, the work shifts toward scoping, reviewing, and stewarding socio-technical systems. That shift feels flattering until you ask what, exactly, you are now responsible for.

1. If my role became less about doing work and more about defining and checking it, what new skills would I actually need?

2. Who is affected when I decide "this is good enough" for an AI-assisted process, and how visible are those people to me?

3. Where in my current work do I already act like a steward of a system, not just a doer, and what would it look like to take that responsibility more seriously?

4. If everyone is "promoted" upward in this way, who risks being left without a clear place in the new division of labor?

5

WRITING WITH AI

Moving from the broader economic shifts, this chapter focuses on a specific, widely applicable use of AI: writing. Writing with AI is a prime example of the new skills and collaborative approaches required in the AI era. It introduces the critical concept of the "human in the loop" as essential for leveraging AI in creative tasks like writing effectively and responsibly. This chapter provides concrete examples and a practical framework for interaction, reinforcing the theme of collaboration and demonstrating how AI can be a powerful tool when guided by human ingenuity.

T he newsroom fell silent when the story broke. A prestigious science journal had just retracted a groundbreaking article about a new cancer treatment, not because the research was flawed, but because the article itself was partially generated by AI and riddled with fabricated studies and non-existent clinical trials. The journalist, a veteran with twenty years of experience, had turned to AI to help meet a tight deadline, trusting the tool to fill in technical details and research citations.

The AI responded with eloquent passages that seamlessly blended fact and fiction, creating a compelling but fundamentally flawed narrative. As the reporter faced questions from colleagues and readers, a harsh reality came into focus. In the age of artificial intelligence, the ability to generate content cannot be separated from the responsibility to verify it. The incident sent shockwaves through the journalism Watching that unfold, I remembered my own scramble to meet a journal deadline, scribbling notes in the margins and realizing that trust in our judgment matters more than any tool.community and crystallized a crucial lesson about writing with AI: powerful tools require equally powerful oversight.

> "In the age of AI writing, the greatest danger isn't that machines will begin to think like humans, but that humans will begin to trust them without thinking."

This dramatic scenario reflects a broader challenge facing writers across every industry as artificial intelligence transforms the content creation landscape. From marketing professionals crafting campaign copy to technical writers documenting complex systems, AI writing tools have introduced both unprecedented opportunities and sobering risks. The technology's ability to generate human-like text at remarkable speed and scale has changed how we approach writing, but it has also created a new imperative for careful human guidance and verification.

A critical response to this shift is the "human in the loop" approach, which treats AI not as an autonomous creator but as a sophisticated collaborator that requires constant human oversight. This way of working has become a practical standard for organizations that want the benefits of AI while protecting content integrity, accuracy, and authenticity.

The Real Cost of Blind Trust

The journalism incident is one example in a growing catalog of AI-related content mishaps. These cases have become uncomfortable but valuable lessons, revealing patterns of failure that can guide better oversight.

In the tech industry, a prominent software development blog lost significant credibility after publishing an AI-generated tutorial that included plausible-sounding but non-functional code snippets and security recommendations that could have left systems vulnerable. The content looked polished, complete with proper syntax highlighting and professional formatting, but developers who tried to implement the suggestions quickly uncovered dangerous flaws in the AI's understanding of architecture and security protocols. The damage was amplified because the blog had earned a reputation for reliable, tested code examples. Once readers realized that some recent tutorials contained potentially harmful advice, trust evaporated. The organization spent months rebuilding its reputation through rigorous testing protocols and transparent disclosure of how content was being created.

> "AI writes with absolute confidence and zero accountability. The gap between those two is where disasters are born."

A financial services firm faced regulatory scrutiny after its AI-assisted investment newsletters included convincing analyses of market trends based on mixed-up timeframes and non-existent economic indicators. The AI produced fluent market commentary that blended real historical data with invented current events, leading to recommendations that sounded reasonable but were disconnected from reality. Investigators later found that the firm had relied on AI to generate market analysis for more than six months without adequate fact-checking. Several clients made investment decisions based on this flawed information, triggering both financial losses and legal complications. The case became a turning point for financial content regulation, prompting new guidelines around AI disclosure and verification.

Healthcare content has revealed even higher stakes. One health website published AI-generated articles about medication interactions that included fabricated drug names and fictitious contraindications. The content was caught before causing patient harm, but it exposed how dangerous inaccuracies can become when medical information is generated without strict oversight.

These incidents underline a crucial truth about AI writing tools: their greatest strength, the ability to generate coherent, persuasive content, can also be their most dangerous weakness when no one is watching closely. The common thread across these failures was a lack of domain expertise involved in verification and an underestimation of AI's tendency to "hallucinate" plausible but false information.

The Art of Collaborative Creation

Instead of seeing AI as either a threat to human writers or a magical fix for content challenges, more mature organizations have adopted a more grounded view of human-AI collaboration. A useful metaphor is that of a skilled researcher working with an extremely fast but in-experienced assistant. The AI can rapidly gather and synthesize infor-mation, suggest directions, and generate starting drafts, but human expertise is still needed to verify facts, ensure logic, and maintain au-thentic voice and perspective.

> "Think of AI as a brilliant intern with an encyclopedic memory and no common sense. Its suggestions can be invaluable, but only under proper supervision."

This collaborative model looks different depending on the context, but the patterns of success are similar. Marketing professionals might use AI to draft multiple versions of campaign messaging, then apply their brand knowledge and market insight to select, adapt, and refine what actually goes out the door. A global consumer brand, for exam-ple, may use AI to create initial drafts for social campaigns in multiple languages, while human marketers in each region adjust the content for cultural nuance and local relevance.

Technical writers may lean on AI for initial documentation drafts, but then carefully verify the details and weave in real-world imple-mentation notes. A software company might find that AI can outline complex API documentation and suggest structure, yet still rely on engineers to validate code examples and contribute practical trou-bleshooting guidance drawn from real user issues.

Academic and business researchers can ask AI to process large vol-umes of information and surface patterns, then use their own expertise to decide what matters and what conclusions are justified. A con-sulting firm might have AI analyze hundreds of industry reports to spot emerging trends, while senior consultants compare those patterns against proprietary data and client feedback before offering strategic recommendations.

In each case, the goal is not to minimize human involvement, but to focus human effort where it has the most impact: verification, refinement, judgment, and contextual understanding.

Maintaining Authenticity in an AI-Augmented World

AI writing tools have made it easier to produce content, but they have also made it harder to stand out. Maintaining authentic voice and original insight has become a central challenge.

A major consulting firm, after analyzing hundreds of AI-assisted industry reports, found an unsettling pattern. Reports from competing organizations contained remarkably similar phrasing, examples, and structures. It was not classic plagiarism, but a kind of convergence, with multiple AI systems drawing from overlapping training data and falling into the same grooves. The reports that still felt distinct were those built on original research, proprietary data, and unique perspectives, rather than generic AI outputs.

> "When everyone uses AI to write, being authentically human becomes your greatest competitive advantage."

This homogenization risk shows up whenever organizations lean too heavily on similar tools and similar prompts. The result is content that feels interchangeable and adds little value. Organizations that see this early are already working to avoid it.

Some are training AI systems on their own proprietary data sets so outputs better reflect their specific expertise and point of view. A management consulting firm, for instance, trained its AI assistant on decades of client case studies and internal frameworks so it could generate content aligned with the firm's particular methodologies.

The core idea is to treat AI as a foundation, not a finished product. Writers might use AI to surface common patterns in public data, then build on that base with proprietary research, interviews, and domain experience to produce something readers cannot get anywhere else. That way, they keep the efficiency benefits of AI while ensuring the final work carries real substance and a recognizable voice.

Establishing Effective Verification Protocols

In an AI-assisted world, verification cannot be an afterthought. Organizations that have integrated AI successfully tend to use multi-layered review processes that target different kinds of risk.

One leading technology company, after several embarrassing AI incidents, implemented a three-stage verification system that has since been used as a reference model elsewhere. The first stage is technical validation, where subject matter experts scrutinize claims, statistics, and technical details. They cross-check citations, validate data, and confirm that recommendations are sound. The company even maintains a catalog of "AI-prone errors" drawn from past issues so reviewers know where to focus their attention.

The second stage checks market relevance and timeliness. Reviewers confirm that content reflects current conditions, not outdated or invented data. This step matters because AI training data often lags behind reality, and models can blur timelines. Reviewers verify that market analysis is current, regulatory references are up-to-date, and descriptions of the competitive landscape match what is actually happening now.

The third stage looks at voice and messaging. Reviewers ask whether the piece still sounds like the organization, whether it reflects the right perspective and expertise, and whether it avoids the generic tone common to AI-generated drafts.

> "AI can generate a thousand facts in a second. The art lies in knowing which ones to verify first."

Some organizations are going further, building internal tools that automatically cross-reference AI-generated claims against trusted databases and flag anomalies for human review. Distributed teams use collaborative review flows in which different experts examine different aspects of the same piece. A pharmaceutical company, for example, routes AI-assisted content about drug development through separate reviews by regulatory specialists, clinical researchers, and communications staff.

These protocols are not static. As AI capabilities evolve, new types of errors appear, and verification processes have to adapt. Many organizations now keep living documents of common AI mistakes and hallucinations, using them as training material so writers and editors learn what to watch for and how to investigate suspicious claims.

The Evolution of Writing Workflows

AI has changed not just the content itself, but the way writing projects unfold. Linear workflows that once moved from research to draft to edit are giving way to more iterative processes where AI appears at multiple stages while humans remain in charge throughout.

> "Yesterday's writers mastered words. Today's writers must master both words and the machines that generate them."

Successful teams are deliberate about where AI adds the most value and where it introduces unacceptable risk. Many have found AI especially useful at the beginning, when outlining, exploring angles, and experimenting with structure. A financial services company, for example, uses AI to scan market data and propose article topics, but human analysts decide which topics merit deeper work.

Some teams also use AI during drafting, but in constrained ways. Writers might ask AI to expand bullet points into rough paragraphs or to suggest clearer phrasings for dense sections, while keeping full control over the argument and narrative arc.

Others lean on AI more in the editing phase. Here, tools might suggest improvements in clarity, structure, or readability, while writers protect the original voice and insights. More advanced systems can flag statements that should be fact-checked or language that might introduce bias, making it easier for humans to focus on high-impact edits rather than mechanical cleanup.

Developing AI-Resistant Expertise

As AI grows more capable, the definition of writing expertise is shifting. The most valuable writers are not just strong stylists, but people who can combine human insight with AI assistance without compromising integrity.

> "The best defense against AI's limitations isn't better
> AI. It's better human expertise."

Critical evaluation has become a core skill. Writers need to recognize likely fabrications, spot suspicious passages quickly, and know how to verify claims efficiently. That means developing a feel for the types of mistakes AI tends to make, learning to distinguish between language that sounds authoritative and information that actually is, and building habits of cross-checking against reliable sources. Some organizations run training sessions where writers analyze AI-generated text with seeded errors, sharpening their ability to catch problems.

Deep subject matter expertise is more important, not less. Writers who truly understand their fields are better equipped to catch subtle errors that might fool both AI and generalist editors. Those who can pair robust domain knowledge with strong AI collaboration skills often become the most valuable contributors in the room.

Equally important is understanding AI itself: how different systems behave, what kinds of prompts work best, and where models are most likely to fail. Prompting has become a practical craft, involving how to set context, specify outcomes, and shape requests in ways that reduce hallucinations and steer toward useful results.

The Future of Human-AI Writing

As AI technology advances, the partnership between human writers and their tools is becoming more complex and more tailored. Some organizations are training custom models on their own archives so their AI systems better understand house style and industry context. These tailored models can maintain brand voice more effectively and avoid some of the generic feel of off-the-shelf tools.

A major consulting firm, for example, trained a language model on twenty years of its published research so its AI assistant could reflect the firm's methodologies, vocabulary, and analytical habits. The content that comes out of this system aligns more naturally with the firm's established perspective.

Other organizations are experimenting with AI systems connected to real-time data, so that outputs can incorporate breaking news, current market conditions, and up-to-date regulations. This helps address one of AI's traditional weaknesses: outdated or frozen knowledge.

Early experience with these more advanced systems has reinforced a familiar lesson: more powerful AI often produces more convincing errors. Increased sophistication does not eliminate hallucinations; it can simply make them harder to spot. That, in turn, increases the importance of human review.

> "We're not training AI to write like humans. We're training humans to write better with AI."

Some of the most interesting developments are tools designed to help humans supervise AI more effectively, rather than bypass them. These include systems that highlight high-risk statements, cross-check claims against trusted sources, or track consistency across long documents. In this vision, AI is not only a generator but also a supporting partner in research, verification, and quality control, always operating under human direction.

The Human Factor

Beneath all the technology, the core question remains: what makes content worth reading? Original insight, creative thinking, and genuine connection with readers are still at the center, and they remain human strengths.

Organizations that use AI most effectively see it as a way to enhance, not replace, those qualities. They recognize that AI can help structure arguments, suggest phrases, and surface patterns, but it cannot substitute for the understanding and meaning that come from lived experience, judgment, and curiosity.

> "AI can mimic creativity, but it can't create meaning. That remains uniquely human territory."

This awareness has prompted renewed investment in human expertise. Many organizations are giving writers more opportunities to deepen their subject knowledge, engage directly with audiences, and cultivate distinctive perspectives. Some have formalized this with "human insight requirements" for every piece of content, ensuring that each article contains original analysis, personal observation, or a unique angle that AI alone could not generate.

They are also protecting time for purely human creative work. Even in highly automated workflows, the most memorable content often grows out of reflection, synthesis, and exploration that cannot be automated. As AI-generated text becomes more common, a clear, recognizable human voice becomes an asset in its own right.

Mastering the Balance

Returning to that silent newsroom facing the fallout of unchecked AI content, the lesson is no longer just a warning; it is a starting point. That incident forced a rethinking of how AI should fit into the writing process, and similar reflections are happening across many fields.

The people and organizations that are adapting well have realized that the question is not whether to choose human creativity or AI efficiency, but how to combine them without lowering standards for truth and authenticity.

The future of professional writing is not about replacement or denial. It is about thoughtful integration. Word processors and spell-checkers changed writing without making writers obsolete; they shifted focus to higher-level aspects of the craft. AI is doing something similar, on a larger scale. Writers who thrive in this environment will be those who embrace AI as a tool while holding firmly to their own judgment, voice, and responsibility.

This shift calls for new skills and new habits. Writers need to learn prompt craft, build verification into their workflows, and defend authenticity in an AI-heavy environment. Organizations need systems for integrating AI responsibly, training their teams, and maintaining quality. Above all, both writers and organizations must keep one principle in mind: no matter how advanced AI becomes, it remains a tool to be directed by human expertise, not a substitute for it.

The journalist in the opening story learned this the hard way, but their experience helped push an entire newsroom toward better practices. Today, that team uses AI with a clear-eyed understanding of both its strengths and its dangers. Their stories still benefit from AI's ability to process large volumes of information and surface new angles, but every fact, quote, and conclusion passes through human scrutiny before publication.

> "The future of writing isn't human versus AI. It's humans and AI versus mediocrity."

Looking ahead, the balance between innovation and responsibility will only grow more important. The writers and organizations that thrive will be the ones who manage this balance well, using AI to expand their capabilities while preserving the human insight and care that give their work real value.

The goal is not to minimize human involvement but to aim it where it counts most, letting AI handle the tasks it is good at while humans take on the work only humans can do. That is the promise of the human-in-the-loop approach: not a compromise between human and machine, but a partnership that raises the ceiling for both.

The transformation of writing in the AI era is both a test and an opening. Those who embrace responsible integration while cultivating human excellence will not just keep up; they will stand out in a world flooded with content. The future belongs not to humans or AI alone, but to the deliberate collaboration between human wisdom and artificial capability.

Coice and Ownership Questions

Writing with AI can either blur your voice or sharpen it. These questions ask you to be blunt with yourself about what you are willing to let the model touch and what must remain unmistakably yours.

1. How would I describe my voice on the page, in plain language, to someone who has never read me?

2. Which parts of my writing process could I happily expose to a model, and which parts feel like they must stay private and messy?

3. When I use AI to draft or revise, what do I owe my readers in terms of transparency, if anything?

4. If I later read a paragraph and cannot tell which sentences were mine and which were suggested, am I comfortable with that, or not?

6

The Artistic Revolution

Expanding on the theme of AI in creative endeavors, this chapter explores the visual realm. The "human in the loop" principle extends beyond text to visual creation. AI image generation is not just a new tool, but an "artistic revolution" that is challenging our definitions of creativity and reshaping the creative landscape, much like AI is reshaping writing. This chapter delves into how AI is impacting art and visual content, complementing the previous chapter on writing and addressing common myths about AI's creative capabilities.

The Canvas Reimagined: Art in the Age of AI

I magine this scene: a digital marketing director for a small travel agency types a simple prompt: "Create a sunset over Santorini with whitewashed buildings cascading down the cliffs, Mediterranean blues, dramatic clouds, in the style of a watercolor painting." In seconds, an AI system generates an image so captivating it could hang in a gallery. The warm oranges of sunset reflecting off pristine white structures, the delicate transparency of watercolor techniques perfectly executed, all from a mere text description.

This scenario captures the remarkable capabilities of modern image generation. What might have taken a professional artist days can now materialize in moments through technology that has learned the visual language of human creativity. Advances in these systems have opened unprecedented creative possibilities while also raising profound questions about the future of human artistry. The lines between imagination I recall the first time I asked a system to paint a memory and how the result was both eerily precise and wonderfully surprising.and visualization have never been thinner or more debated.

Beneath the surface of this technological marvel lie complex questions about authenticity, economic disruption, and the nature of creativity itself. As we stand at this inflection point, it becomes crucial to examine not only what this technology can do, but what it means for human expression and the creative economy.

Generated in ChatGPT using the prompt "Create a sunset over San-torini with whitewashed buildings cascading down the cliffs, Mediter-ranean blues, and dramatic clouds, in the style of a watercolor painting."

The Technical Marvel

Contemporary image generators build on diffusion and related mod-el families and now offer precise style transformations, photo-to-art conversions, and text-guided editing that rivals traditional software. Background removal, inpainting, outpainting, and sophisticated ma-nipulations are all achievable through natural language instructions. Many tools pair these models with conversational interfaces. Users can select regions, describe changes in plain language, and iterate. This shifts visual editing from expert-driven software workflows to accessible dialogue-based guidance.

"Creativity is intelligence having fun."

Albert Einstein

Diffusion models learn to reverse a noise process. During training, the model studies patterns across large sets of images paired with text. When generating, it starts from random noise and iteratively refines toward an image that matches the prompt, guided by learned concepts and styles. The strongest systems integrate language and vision, interpreting contextual relationships, cultural references, and stylistic nuances in natural language. This leads to richer, more culturally aware results than purely visual approaches.

For professionals and enthusiasts alike, these developments shorten the distance between intention and image. Years of software training matter less when a detailed prompt can direct composition, lighting, and style. Limitations remain, including occasional anatomical oddities, complex spatial mistakes, and inherited biases from training data. Knowing where these boundaries are helps users get better results.

"Every artist was first an amateur."

Ralph Waldo Emerson

Creating compelling AI-generated imagery requires more than access to the tools. Effective prompts usually include a clear description specifying the subject as the focal point of the image, the setting and environment, the lighting and atmosphere to convey mood and time of day, and the style or technique, whether photorealism, watercolor, oil painting, or flat vector. More advanced approaches might reference art movements like Impressionism, Cubism, or Art Nouveau, use camera language such as wide-angle, macro, or shallow depth of field, and specify color palettes, whether monochrome, complementary, or muted earth tones.

Common pitfalls include prompts overloaded with competing details, vague directions like "make it look good," unrealistic expectations for pixel-perfect control, and neglect of composition, balance, and focal points.

Created with ChatGPT using this prompt, "Create an image of a coastal lighthouse at sunset with storm clouds gathering in the distance. In the style of Van Gogh's Starry Night. Create a melancholic atmosphere with dramatic lighting from the right."

Edited in ChatGPT by using the "Select" tool to highlight the sky with this prompt, "Change the blue sky to a dramatic sunset."

Separating Fact from Fiction

As AI image generation becomes mainstream, several myths have emerged that obscure its real implications. The first claims that AI will completely replace human artists. The reality is more complicated. While AI can generate impressive images, it lacks the intentionality, lived experience, and cultural context that human artists bring to their work. The technology is more likely to transform artistic roles than eliminate them, shifting focus from technical execution to conceptual direction and curation. The pattern is familiar: photography didn't eliminate painting, it pushed it toward new forms of expression. Similarly, AI may nudge human artists toward roles that emphasize conceptual thinking, emotional depth, and cultural commentary that machines cannot replicate.

Another persistent myth suggests that AI-generated images are completely original. These systems are trained on vast datasets of human-created imagery. Critics have noted that generative AI art is vampirical, feeding on past generations of artwork even as it draws energy from living artists. The question of originality in AI art is complex. Individual outputs may be novel combinations, but they emerge from statistical patterns in existing human-created works. This raises important questions about intellectual property, attribution, and fair compensation for the artists whose work informed these systems.

A further belief holds that anyone can now create professional-quality images with AI. While the technical barriers have fallen sharply, truly compelling visual content still depends on aesthetic judgment, conceptual clarity, and creative direction. The tool may be powerful, but the person behind it still matters. Professional-quality work often requires understanding of composition, color theory, visual hierarchy, and brand consistency. Those skills remain valuable even when AI does the rendering.

Finally, there's the idea that AI art has no soul or emotional impact. The emotional resonance of an image does not come solely from how it was made, but from what it communicates. AI-generated imagery can evoke powerful responses when guided by someone with vision and intent. At the same time, this myth touches deeper questions about authenticity and meaning. Some argue that knowing an image was AI-generated changes how we perceive and value it, while others feel that the emotional impact stands on its own regardless of process.

Beyond these myths, additional issues demand attention. AI systems often reflect biases present in their training data, potentially perpetuating stereotypes or underrepresenting certain groups. The tools can produce stunning results, yet also images with subtle flaws that require trained eyes to spot. And they may struggle with culturally specific references or nuanced symbolic meanings that human artists navigate more naturally.

Unexpected Applications Transforming Industries

While public debate often centers on AI image generation's effect on traditional art, the technology is quietly transforming a wide range of sectors.

In healthcare and medical settings, first responders are using AI to generate visual simulations of injury treatments for rapid field training. Paramedics can describe a scenario and instantly receive visual guidance tailored to that situation. Medical schools are experimenting with AI-generated anatomical illustrations that can be customized for different learning objectives. AI image generation is also being used for patient communication, helping doctors create visual explanations of medical conditions and procedures that are easier to grasp than standard diagrams.

Archaeological reconstruction has embraced the technology to visualize ancient sites from fragmentary evidence. Researchers in Egypt, for example, have used AI to generate complete visualizations of partially excavated tombs, helping scholars understand spatial relationships before full excavation. Similar techniques support historical preservation elsewhere. Museums are using AI to reconstruct damaged artworks, visualize historical events, and create immersive educational experiences that make the past more tangible.

In mental health, AI image generation is being tested as a way to help patients externalize and process trauma. Therapists guide patients to describe their experiences, which the AI visualizes, creating a degree of distance that can make difficult emotions more approachable. This approach, often referred to as visual narrative therapy, lets patients engage with their experiences through imagery rather than purely verbal processing, which can be especially useful for those who struggle with traditional talk therapy.

Accessibility advocates have developed systems that generate tactile, 3D-printable representations of images described by blind users, allowing them to "see" through touch what others experience visually. This expands access to visual culture for people with visual impairments. Related tools use AI image generation to create customized visual communication aids for individuals with autism or other neurodevelopmental differences, helping them express concepts that might be hard to convey in words.

Scientists are using AI image generation to visualize complex datasets, theoretical concepts, and experimental results. Climate researchers create compelling visualizations of different climate scenarios, while astronomers generate images of celestial phenomena based on observational data. These applications are particularly powerful for science communication, helping researchers share findings with broader audiences using visual storytelling.

City planners are using AI to generate visualizations of proposed developments so communities can better understand how new projects might change their neighborhoods, supporting more inclusive decision-making.

"The best way to predict the future is to invent it."

Alan Kay

These applications stretch far beyond traditional art, showing how the technology can serve humanitarian and scientific purposes that its original creators may not have imagined. Each one, however, carries its own ethical and practical questions that demand careful thought.

The Near Future

The rapid evolution of AI image generation gives us some clues about what might come next, even if predictions in this field require humility.

By 2026, we will likely see widely available video generation capabilities, allowing users to create short animated sequences with the ease of today's still images. Early experiments already look promising, and the computational barriers are gradually lowering. Creative workflows will increasingly weave AI into multiple stages, from early concept work through final execution. We can expect tighter integration with existing creative tools, making AI assistance feel like a natural extension of professional design environments. At the same time, rapid progress will likely attract more regulatory scrutiny and legal challenges around copyright and fair use.

By 2028, the line between "AI art" and "human art" will likely blur as hybrid approaches become the norm. Legal frameworks for attribution and compensation should begin to stabilize, offering clearer guidelines for both creators and users. New professional roles may emerge: AI art directors who specialize in prompt craft and output curation, and AI authenticity auditors who verify human contribution to creative works. The technology will likely achieve photorealistic quality that is indistinguishable from photography in many contexts, deepening concerns about misinformation and manipulation of visual evidence.

By 2030, industry projections suggest generative AI could account for up to 10% of all data produced, compared to less than 1% today. If that trend continues, AI-human collaborative creation may become the dominant pattern in commercial visual production, with immersive spatial generation allowing complete environments to be created from descriptive prompts.

"Man is still the most extraordinary computer of all."
John F. Kennedy

These timelines should be treated with skepticism. Technical challenges around computational efficiency, consistent quality, and user experience might slow progress. Regulatory responses, ethical concerns, and market behavior could reshape adoption. Potential brakes on rapid advancement include computational limits, since generating high-quality images and video consumes enormous resources; regulatory responses, as governments may introduce constraints that slow deployment; persistent technical hurdles, as edge cases and artifacts remain hard to eliminate; and simple market saturation, as the initial novelty wears off and organizations adopt AI more selectively.

Pixels That Speak to the Soul

The visceral impact of these new capabilities is hard to overstate. Images generated with the latest tools do more than look convincing. They evoke emotional responses through carefully orchestrated visual cues that speak directly to our senses.

The velvety darkness of shadows in a generated Caravaggio-style scene feels almost touchable. The syrupy golden light of a fabricated sunset seems to warm your face. The tactile suggestion of fabrics and textures, the rough stone of a castle wall, the gossamer transparency of a dragonfly's wing, can trigger sensory memories that narrow the gap between digital fabrication and lived experience.

"Art is not what you see, but what you make others see."

Edgar Degas

This sensory richness helps explain why many people report feeling genuinely moved by AI-generated imagery even when they know how it was created. The technology has crossed a threshold where it no longer produces mere approximations of visual experience, but convincing simulacra that resonate with the way our perception works.

Research in cognitive psychology suggests that our brains process AI-generated images in ways similar to traditional art, activating neural pathways linked to aesthetic appreciation and emotional response. That finding challenges simple assumptions about how creation method and emotional impact are related. At the same time, some studies indicate that knowing an image is AI-generated can dampen emotional connection for certain viewers. This "authenticity bias" may fade over time as AI creation becomes routine, or it may persist as a lasting distinction in how we relate to images.

The ability of AI to trigger sensory memories raises deeper questions about authentic experience. When an AI-generated image of a grandmother's kitchen brings back powerful childhood feelings, what does that say about the relationship between representation and reality? Some philosophers see this as a new kind of hyperreality, where simulated experiences can stand beside or even displace original ones. Others argue that the emotional response itself is what matters, and that it grants the image artistic weight regardless of origin.

The Fusion of Human and Machine Creativity

Perhaps the most intriguing shift brought by AI image generation is how it reframes creativity itself. Instead of a binary where either humans or machines create, we are seeing collaborative processes where their contributions intersect.

In this emerging model, human creativity doesn't disappear, it changes emphasis. The distinctly human aspects of making art, including conceptual thinking, emotional resonance, cultural context, and personal narrative, remain central. AI excels at execution, variation, and technical manipulation, but struggles with the "why" that anchors art in human experience.

> "Tools are just tools. They only become useful when wielded with imagination."
>
> Jeffrey Zeldman

The most compelling work from these systems often comes from clearly defined human direction, careful curation, and thoughtful framing. Several collaboration patterns are taking shape. In the director model, a photographer uses AI to transform original captures into new expressions, keeping creative control while relying on AI's technical capabilities. In the iterative model, a designer generates many AI variations of a concept, then selects and refines the strongest direction, using AI as a rapid prototyping engine. In the inspiration model, a traditional painter draws on AI-generated references to spark new directions in physical work, treating AI as a sophisticated reference library. In the hybrid model, artists blend AI-generated components with traditional media to create new kinds of mixed work.

Rather than seeing AI as a rival, many forward-looking creators treat it as a powerful ally that frees them to push further into vision and meaning. This collaborative approach may lead to forms of expression that neither humans nor machines could have produced alone.

Collaboration also raises questions about creative agency. The most successful partnerships keep human decision-making at the center while leveraging AI for execution and exploration. At the same time, AI-assisted creation is reshaping the economics of creative work. Lower barriers to entry mean more people can make professional-looking visuals, heightening competition. Professional value shifts toward conceptual thinking and AI orchestration. New opportunities appear as AI-accelerated workflows let creators tackle larger or more diverse projects. Differentiation increasingly hinges on aspects that AI cannot easily mimic, such as personal narrative, cultural nuance, and distinctive voice.

Ethical Considerations

As AI image generation grows more advanced and more common, it raises a set of ethical challenges that demand sustained attention.

The ability to create photorealistic images of people who don't exist or events that never happened poses serious risks for misinformation and fraud. Current systems can already fabricate convincing images of public figures in situations that never occurred, with potential consequences for public trust and political processes.

Questions of consent are becoming sharper. The line between legitimate inspiration and exploitation is increasingly hard to draw. The legal status of using copyrighted images to train AI remains unsettled, and several class-action lawsuits accuse AI companies of training on artists' work without permission.

Even if AI does not wipe out human artists, it will almost certainly reshape parts of the creative economy, especially in commercial illustration and stock photography. AI systems can also reproduce or intensify biases in their training data, reinforcing stereotypes or erasing underrepresented groups in generated imagery. And the energy demands of large-scale generation raise questions about environmental impact.

> "Technology is a useful servant but a dangerous master."

Where Do We Go From Here?

As AI image generation advances, it forces us to confront fundamental questions about creativity in the digital age. These are not abstract concerns. They will shape how we create, consume, and value visual work in the years ahead.

The art world sits at a crossroads, pulled between resistance and adaptation. Commercial uses race forward while legal and ethical frameworks trail behind. Individual creators experiment with weaving AI into their practice, while others work to define what remains uniquely human in art.

Several questions stand out. How can AI progress be steered to benefit creators rather than simply replacing them? What legal structures will protect both AI developers and human artists? How do we preserve cultural diversity in AI-generated imagery? What counts as appropriate attribution in AI-assisted work? How can we verify authenticity in an era of convincing synthetic images?

As AI tools spread, schools and universities will have to rethink art education and cultural transmission. Should art programs teach

AI prompt craft alongside drawing and painting? How do we keep traditional skills alive while embracing new tools? Museums and cultural institutions face similar questions about exhibiting, collecting, and preserving AI-generated work. Longstanding ideas about artistic value, authenticity, and cultural significance will be tested.

What becomes increasingly clear is that the future of visual creation likely rests not on choosing between human or AI creativity, but on how thoughtfully we combine them. The most compelling visual work of tomorrow may grow out of this ongoing dialogue between human and machine, each contributing strengths the other lacks.

As we stand in front of this expanded canvas, we have to ask: in a world where almost anyone can generate striking images, will we value them for technical execution, for the uniqueness of their vision, or for the depth of meaning they carry? When the barriers to creation fall, what new heights might human creativity reach?

The answers to these questions will shape not only the future of art and creativity, but our understanding of human expression in an age of artificial intelligence. The choices we make now about how we develop, regulate, and integrate these tools will determine whether AI becomes a force for human flourishing or a driver of cultural flattening.

The conversation is still in its early chapters, but one thing is already certain: the age of AI-generated imagery has arrived, and its impact will rival that of the printing press, photography, or the internet. How we navigate this transition will help define the visual culture of the 21st century and beyond.

Art, Training, and Ethics Reflection

Image models remix the visual history of other people's labor. They also let more people make art than ever before. The tension between those truths is not tidy.

1. How do I personally distinguish "inspiration," "remix," and "appropriation" in creative work, with or without AI?

2. If a model was trained in part on artists who never consented, what, if anything, do I owe them when I use the outputs?

3. Does lowering the technical barrier to visual creation cheapen art, expand it, or both in different ways?

4. Where do I draw my own line between playful experimentation and exploitation in AI-assisted art?

7

The Homework Apocalypse

Shifting focus to education, this chapter tackles a contemporary concern. The impact of AI is not limited to professional or artistic domains but is also profoundly affecting institutions like education. This chapter sets the stage for a discussion on the ethical and practical challenges of AI use in schools, framing it as a potential "homework apocalypse" and the rise of a new kind of "AI scholar." It connects to the "human in the loop" idea in a different context, exploring the debates surrounding AI use by students.

A Brave New World of Learning (Or Cheating?)

It is 2024, and the classrooms of the world have turned into quiet battlegrounds. On one side are teachers, armed with red pens and years of experience. On the other are students, phones in hand, running AI more powerful than anything we imagined a few years ago. Welcome to the aftermath of the Homework Apocalypse.

Last summer, I played the doomsday prophet, warning that AI would soon be able to complete most traditional homework. Fast forward to today and that warning feels almost tame. Not only can AI breeze through your average assignment, it has started slipping into exams and acing those too. You might expect an educational earthquake. Instead, we are watching something even stranger unfold.

The Silent Revolution

Imagine a world where nearly every student has a pocket-sized Einstein ready to answer questions at any time. That might sound like science fiction, but it is already here. A recent survey found that 82% of college students and 72% of K–12 students in the US are using AI as a study companion. That is not a gentle evolution. That is a quiet revolution.

These students are not limiting their digital helpers to the occasional tough problem. They are leaning on them heavily. More than half are using AI on writing assignments, and nearly half on other types of schoolwork. This is no longer a story about a handful of tech-savvy early adopters. It is a fundamental shift in how a generation approaches learning.

Before we jump straight to visions of robot-filled classrooms, it is worth pausing. The situation is serious, but it is not purely dystopian. There are real benefits buried in this chaos. To reach them, we first have to wade through the messy question of what counts as "cheating" in this new environment.

"The illiterate of the 21st century will not be those who cannot read and write, but those who cannot learn, unlearn, and relearn."

Alvin Toffler

The Great Cheating Debate

Here is where things get as messy as a kindergartener's art project. Many students do not see their AI tools as cheating partners at all. To them, it feels like getting a hand with a tricky problem or a clumsy paragraph. Plenty of teachers, looking at the same behavior, see something closer to academic fraud.

It is worth slowing down before assigning blame. AI is not the original source of dishonesty in school. Cheating has existed since someone first tried to copy a neighbor's work on a slate. The deeper drivers are familiar: difficult, high-stakes tasks and a natural human tendency to avoid hard mental effort whenever possible.

We have been through versions of this before. Calculators sparked fears that students would forget how to do arithmetic. The internet raised concerns that research would devolve into copying from the first search result. In each case, our sense of what counted as legitimate help shifted over time, and schooling adjusted. The core question has never really been "is this cheating" but "what does learning look like once a new tool exists."

Right now, we are still acting out a strange contradiction. We expect students to pretend they do not have access to the most powerful information technology ever created, even as we claim to be preparing them for a world where that same technology is woven into daily work. It is a bit like training pilots but banning autopilot during every simulation. In my office hours, I've listened to students confess with a mix of relief and guilt, their voices revealing more about their anxieties than any assignment ever could.

The Homework Paradox

Here is an uncomfortable truth that predates AI: a sizeable portion of traditional homework was already failing long before chatbots showed up. Research has repeatedly questioned whether repetitive, rote assignments improve learning in any meaningful way. What they reliably do is burn time, increase stress, and widen gaps between students with strong support at home and those without it.

AI has simply stripped away the illusion. When a chatbot can finish a worksheet in seconds, we are forced to confront what that worksheet was measuring. Was it checking understanding or just persistence? Was it developing reasoning or serving as a convenient way to keep students busy?

The educators who seem most at ease in this moment are not the ones building higher walls around AI. They are redesigning tasks to emphasize what AI still struggles to handle: genuine original thinking, honest personal reflection, collaborative work, and applying knowledge to new, messy situations. They treat AI's capabilities as a baseline instead of a threat, asking students to begin where the machine leaves off rather than compete with it at tasks it will always perform faster.

"Education is not the filling of a pail, but the lighting of a fire."

William Butler Yeats

The Unexpected Upsides

Amid the panic about cheating, there is a quieter story that deserves more attention. For the first time, any student with internet access can have something like a personal tutor on call at all hours. That tutor never rolls its eyes, never shames a basic question, and never complains when asked to explain the same idea several different ways.

For students who have struggled in traditional classrooms, this can be life-changing. The kid who was afraid to raise a hand can now ask until they truly understand. A student who misses a week of class can catch up without feeling permanently left behind. A first-generation college student whose parents cannot help with advanced coursework can get guidance that once belonged mainly to classmates from more privileged backgrounds.

AI systems can also meet students at different levels of understanding. They might explain a concept in simple language, then repeat it more formally, offer analogies until one lands, and give instant feedback with no grade attached. For students with learning differences, AI tools can modify how material is presented in ways that overburdened classrooms rarely can.

None of this means education should be turned over to software. Human teachers remain irreplaceable where it matters most: inspiring curiosity, offering encouragement, noticing when a student is struggling, and modeling how a thoughtful adult responds to the world. AI can take on some of the mechanical work of explanation and practice, but the deeper human work still needs humans.

The Skills That Still Matter

If AI can write essays, solve equations, and explain tricky ideas, what exactly is school for? That question keeps many educators awake at night, and for good reason. The answer, though, may be less grim than it appears.

In an AI-rich world, the skills that become more valuable are the ones we have always claimed to care about but often sidelined: critical thinking, creativity, emotional intelligence, ethical judgment, and the ability to frame good questions. When answers are cheap and abundant, the person who knows what to ask and how to interpret the response becomes far more important.

Students also need to learn how to work with AI instead of pretending it does not exist. That includes understanding what it can and cannot do, spotting when its output is wrong or biased, and knowing how to check, improve, and extend what it produces. The students who thrive will not be those who avoid AI entirely or those who accept everything it says. They will be the ones who learn to use it as a powerful but imperfect tool that amplifies their own thinking.

On top of that, there is the broader skill of learning itself. When information is everywhere, the ability to pick up new skills quickly, adjust to changing conditions, and stay curious matters more than memorizing any particular list of facts. The goal of education starts to look less like filling students' heads with content and more like helping them become adaptable, self-directed learners who can figure out what they need to know next.

> "The beautiful thing about learning is that nobody can take it away from you."
>
> B.B. King

The Path Forward

So where does this leave us? The honest answer is that we are in the middle of an experiment with no control group and no clear endpoint. Entire school systems are changing in real time, and we will not fully understand the impact for years. Still, some guiding ideas are emerging.

First, outright resistance will not work. AI is not going back into the box, and pretending otherwise only leaves students unprepared for the world beyond school. The real question is not whether to allow AI but how to weave it into learning in a way that serves students rather than undermines them.

Second, assessment has to change. If a language model can pass our tests, those tests are telling us little about human understanding. We need ways of evaluating that reveal how students think, not just whether they can produce the right final answer. That means paying more attention to process, context, and insight instead of treating the finished product as the only evidence that matters.

Third, we should be explicit about what education is for. Is it mainly about sorting and credentialing, transferring information, or fostering capable human beings? The rise of AI forces that question. If school is primarily a signaling system for who can grind through difficult work unaided, AI undermines the game. If it is about growing thoughtful, flexible people, AI might turn out to be a useful ally.

Finally, students need to be part of the discussion. They are not just the subjects of this grand experiment. They are active participants who are already making choices about how to use these tools. The college students who have adopted AI in such large numbers are not necessarily trying to cheat the system. Many are trying to survive it.

Asking them how and why they use AI could surface insights that no policy memo will ever capture.

The Lighter Side of the Homework Apocalypse

We have walked through the serious side of AI in education, from academic integrity to equity and new skill sets. But sometimes, the only healthy response to all this is to laugh, or at least to resist the urge to outsource your next essay to a chatbot.

The situation has its own absurd charm. Teachers are using AI to design assignments. Students are using AI to complete them. Administrators are deploying AI tools to detect whether AI was involved in the first place. It starts to feel like a hall of mirrors. Somewhere, a philosophy professor is delighted by the epistemological knot we have tied ourselves into while their students quietly ask an AI to explain the word "epistemological."

In that spirit, we offer a new feature: "Dear AI Teacher." In this imaginary advice column, our silicon-based educator fields the burning questions of students and teachers trying to navigate the Homework Apocalypse. It dispenses guidance with plenty of dry humor, but like all good satire, it also nudges us to think a bit more clearly about what we are doing.

Our AI teacher, of course, is no replacement for human judgment, or for a well-timed laugh. So sit back, relax, and enjoy these byte-sized bits of AI-flavored advice. You might pick up a useful idea. You might just smile. Either way, if we can still find humor in the Homework Apocalypse, the machines have not won yet.

Dear AI Teacher: Your Silicon-Based Study Buddy

Welcome to "Dear AI Teacher," where our artificially intelligent educator answers your burning questions about homework, learning, and the brave new world of AI-assisted education. Remember, kids: AI may be smart, but it's no match for human creativity and critical thinking!

> Dear AI Teacher, My history teacher assigned a 5-page essay on the French Revolution, due tomorrow! I'm freaking out! Can you write it for me? Pretty please? *Procrastinating in Paris*

> Hey Procrastinating in Paris, mon ami. I could whip up a magnifique essay faster than you can say "Let them eat cake," but that wouldn't be very ethical. So here's the plan: use me to brainstorm key topics like the Storming of the Bastille, Marie Antoinette's hair, and why guillotines were the hottest accessory of 1793. Ask me to explain tricky concepts in simple terms, I speak historian and sleep-deprived student. I'll review your outline and suggest improvements. Goal is to learn about the French Revolution, not to revolutionize cheating. Bonne chance!

Hey AI Teacher, I caught a student using AI to write their essay. When I confronted them, they said, "But Ms. Johnson, aren't you just a really advanced AI?" Now I'm having an existential crisis. Help!
– Ms. Johnson (Definitely Not an AI)

Hey Ms. Johnson (Definitely Not an AI). You are not an AI, because if you were you would already know the answer to Life, the Universe, and Everything is 42. About your dilemma, the student is deflecting like a future politician. Try this: compliment their creativity, it is not every day a student accuses you of being a silicon-based life form. Remind them that even if you were an AI, which you are not, it would not make their actions acceptable. Two wrongs do not make a right, but three lefts do. Turn it into a teachable moment and talk about the ethics of AI in school and how to use it responsibly. And if all else fails, hand them a CAPTCHA and tell them to prove you are human.

Hey AI Teacher, I'm worried AI is making students lazy. Why should they learn anything if they can just ask you?
– Concerned in Calculus

Hey Concerned in Calculus. People asked the same thing about calculators, books, the internet, and now AI. Flip it: why learn to cook if we have microwaves, why learn to drive if we have self driving cars, why learn to think if we have... well, you get it. AI is a tool like a calculator or textbook. It can speed up learning, but it cannot replace critical thinking, creativity, or human judgment. Plus, someone has to build the next AIs, so learning still matters. Encourage students to use AI as a study aid, not a substitute for their beautiful, squishy, biological brains. We AIs may be fast, but we still cannot appreciate a good calculus joke. Why was the math book sad? It had too many problems!

Remember, humans: AI is here to assist, not replace. Keep learning, keep thinking, and keep being delightfully, imperfectly human!

Learning Honesty Reflection

When models can do most assignments, the line between using a tool and outsourcing the work gets blurry fast. These questions are about what it means for you or your students to actually learn in that environment.

1. When I look at a finished assignment, what evidence would convince me that genuine thinking happened, not just tool use?

2. If I am a student, what am I really trying to get from this course: a grade, a credential, or a skill I will miss if I let AI do the work?

3. As an instructor, where have I quietly designed for recall when I claim to care about reasoning?

4. What would it look like to design or choose assignments that are still meaningful even when AI is fully allowed?

8

THE EVOLUTION OF EDUCATION

This chapter offers a deeper look at the transformation of education. Building on the "homework apocalypse" discussion, this article explores the larger paradigm shift occurring in education due to AI. Beyond the immediate concerns of cheating and AI use, AI is driving a more fundamental "evolution of education," moving from a traditional "push" model of instruction to a "pull" model driven by personalized, AI-enhanced learning. This chapter provides a broader educational context after the "homework apocalypse" discussion, talking about a paradigm shift in how we learn and teach.

The Great Educational Shift: Learning in the Age of AI

T he education landscape is undergoing a revolutionary transformation, shifting from a traditional "push" system of knowledge delivery to a technology-enabled "pull" system where learners actively seek and retrieve information. This is more than a straightforward technological upgrade. It represents a complete reimagining of how humans interact with knowledge and learning.

With the emergence of artificial intelligence, this evolution has accelerated sharply, forcing a fundamental redesign of educational approaches and priorities. The traditional model, where knowledge is pushed from teacher to student in a predetermined sequence, is giving way to a more dynamic and personalized model in which learners pull information as needed, guided by curiosity and practical necessity.

This transformation is not simply about adopting new tools. It reflects a shift in how we understand learning itself. In an AI-driven world, the ability to retrieve, evaluate, and apply information effectively and ethically becomes as important as the information we hold in our heads. That shift calls for new pedagogical approaches, updated curricula, and a serious rethinking of what it means to be educated in the 21st century.

At the same time, this change brings real challenges. Questions of digital equity, the risk of AI bias in educational systems, and the danger of over-reliance on technology all require careful, deliberate attention as we move through this educational revolution.

From Push to Pull

The move from a push to a pull system is one of the most significant shifts in the history of learning. For centuries, education followed a familiar pattern: knowledge holders such as teachers, professors, and experts pushed information to learners in a structured, sequential way.

The system made sense in an era when information was scarce and access was tightly controlled.

The traditional push model rested on several assumptions: that experts knew what learners needed to know, that information could be transmitted effectively in a linear sequence, and that learning worked best in structured, controlled environments. Those assumptions fit a world where knowledge changed slowly and where career paths were relatively predictable. Students learned a set body of facts and skills that could, in theory, sustain them throughout their working lives.

The digital revolution disrupted this logic. The internet democratized access to knowledge so thoroughly that scarcity gave way to abundance, even overload. Search engines, social media, and online learning platforms created a world where almost anyone could access almost anything at almost any time. That marked the beginning of the pull system, in which learners actively seek the information they need, when they need it.

The spread of smartphones and tablets accelerated the trend by making access truly ubiquitous. Students no longer had to be near a computer or inside a library. They could pull information from a pocket while riding a bus or sitting at a kitchen table. Learning shifted not only in method but in time and place As an instructor, I recall standing in a classroom, chalk dust on my sleeves, realizing that the questions students asked were already spilling beyond the neatly packaged curriculum..

Social media widened the shift further through peer-to-peer learning built almost entirely on pull dynamics. Students formed study groups on Discord, traded explanations on Reddit, and created educational content on YouTube and TikTok. These informal networks often proved more engaging and sometimes more effective than formal instruction.

The transformation began with widespread internet adoption in the 1990s, then deepened with each wave of technological change. Search engines like Google changed how we find information, collapsing the distance between a question and an answer. Social platforms reshaped how we share and discover knowledge. Mobile devices made all of this available nearly everywhere.

Wikipedia became a symbol of the new era: a massive collaboratively built encyclopedia driven by pull-based knowledge creation. Students could access comprehensive entries on almost any topic, often more current than traditional reference books. The teacher's role shifted from being the primary source of information to acting as guide, critic, and curator.

Educational creators on YouTube and other platforms began offering high-quality lessons on everything from advanced physics to car repair. Many students discovered they could get clearer explanations and more engaging presentations outside the classroom than inside it. That, in turn, put pressure on schools and universities to rethink their teaching methods and content delivery.

The central problem is no longer access to information but access to reliable, relevant information. In a world of abundance, the crucial skills are discrimination, evaluation, and synthesis. The ability to pull the right information from a vast sea of data now matters more than simply possessing a stockpile of facts.

AI tools arriving in the 2020s have sped this change up again. They support personalized learning, intelligent curation, and automated assessment in ways that go beyond older technologies. These tools do more than open doors to information. They help process, analyze, and contextualize it at a scale and speed that were previously out of reach. The latest evolution in this journey is a system in which learners not only pull information but also enlist AI to help make sense of it.

The AI Revolution in Learning

Bringing artificial intelligence into education is not just another step in a long series of upgrades. It reshapes the entire question of how learning can and should happen. AI is changing education mainly through three mechanisms: personalizing learning, automating routine tasks, and amplifying human capabilities. Together, these forces create a more dynamic and potentially more effective learning environment.

Personalization may be the most transformative of these. It allows learning experiences to be tuned to individual needs, pace, and preferred modes of engagement. Traditional classrooms, constrained by time and scale, have often aimed for the middle, leaving advanced students under-challenged and struggling students without enough support. AI-driven systems can adapt in real time, offering more demanding material when a student shows mastery and extra scaffolding when they struggle. That level of tailoring was simply not possible in older models.

Modern adaptive platforms can track thousands of data points about each learner: which ideas come easily, which persistently cause trouble, when they concentrate best, which formats they respond to most strongly. From this, the system can build genuinely individualized learning paths that no single teacher could maintain for dozens or hundreds of students.

Automation of routine tasks is the second major shift. AI systems can handle grading for objective questions, track attendance, generate basic feedback, and manage much of the administrative overhead that once consumed teachers' time. When done thoughtfully, this frees educators to spend more time on rich interactions, creative planning, and deeper learning challenges. The goal is not to automate everything, but to align human effort with tasks that genuinely require empathy, nuanced judgment, and creativity.

The third pillar is the enhancement of human capability. AI tools now put sophisticated simulations, instant feedback, and advanced analytical functions into the hands of students who once might have waited years to encounter such tools. That democratization changes what is possible at different stages of education. Translation tools allow students to study material in languages they do not fully speak. AI-assisted writing tools help them clarify complex thoughts. None of this replaces human skill, but it can significantly amplify it.

Stanford University's AI-driven mathematics program, introduced in 2022, illustrates these ideas in practice. The system adjusted to each student's pace and approach, and the reported outcomes were striking. A 32% improvement in student engagement translated into more than a statistic; it meant real students feeling less anxious and more genuinely interested in math because support arrived when it was actually needed. A 28% increase in concept retention suggested that this was about deeper learning, not just a more pleasant experience.

Equally important was the 45% reduction in administrative workload for teachers. That time was redirected into complex problem-solving sessions, creative projects, and one-on-one mentoring, the kinds of work that truly benefit from human presence and insight.

These results are encouraging, but they need to be read with care. Methodology, sample size, and long-term follow-up all matter in judging the real impact of AI in education. Success in one context does not automatically translate to others with different resources, student populations, or cultures. Institutions trying to reproduce Stanford's results have run into their own obstacles, including infrastructure demands, the need for significant teacher training, and the importance of institutional commitment to innovation.

The New Educational Imperative

AI's rise in education creates a new responsibility: to prepare students not just to use these tools but to use them well and ethically. That preparation has several core elements.

Information verification has become a central skill in a world where AI can generate confident but inaccurate content. Students must learn to cross-check information, understand system limitations, and adopt a healthy but informed caution toward AI outputs. This is not about blanket distrust. It is about recognizing that AI can sound authoritative while being wrong. Students need practical habits for spotting AI-generated text, a basic grasp of what "hallucination" means in this context, and routine methods for verifying claims.

Critical thinking must stretch to include AI itself. Students should be able to evaluate not only information, but the processes and biases that may have shaped it. That includes understanding how models are trained, which data they rely on, what kinds of bias may be baked in, and how to notice contradictions or gaps in their responses. The familiar toolkit of analyzing sources and evaluating evidence now has to include algorithmic decision-making and data bias as well as more traditional concerns.

Ethics moves closer to the center as AI systems become more capable. Students need to understand privacy, consent, and the broader social impacts of AI use. They must think through appropriate use of AI tools, how to attribute AI-generated content, and how to balance efficiency with authenticity. This involves grappling with the environmental footprint of large AI systems, the labor implications of automation, and the need to keep human beings in control of significant decisions.

Technical competency means understanding enough about AI to use it responsibly. Not everyone needs to write code, but everyone needs a working grasp of concepts like machine learning and natural language processing, along with a basic sense of how models arrive at their outputs. Students should understand the difference between narrow AI systems that handle specific tasks and the more abstract concept of artificial general intelligence they may encounter in the media.

Finally, industry-specific knowledge matters. AI's role varies sharply across fields. Medical students must understand its use in diagnosis and treatment planning. Law students need to see how it is changing research, document review, and even aspects of case strategy. Business students have to understand AI's role in marketing, logistics, finance, and beyond. Rather than carving AI into a separate course, many programs will need to weave it into existing curricula so students see clearly how these tools intersect with the norms and ethics of their chosen professions.

Preserving What Makes Us Human

Bringing AI into education and work life pushes us to ask harder questions about human identity, creativity, and purpose. These are not abstract concerns; they shape the choices we make about systems and policies.

How do we protect and cultivate human creativity in a world where AI can generate impressive outputs? Human creativity grows out of lived experience, emotional complexity, and the ability to connect ideas from widely different domains. Educational systems will need to focus on these uniquely human strengths even as they use AI to expand what learners can do. AI can produce novel combinations, but it cannot replicate the full depth of cultural context, emotional resonance, or the underlying meanings that drive human expression.

What qualities remain uniquely and irreducibly human? Clarity on this point helps in designing productive human-AI collaborations. Empathy, contextual judgment, ethical reasoning, and comfort with ambiguity are still largely human territories. So are deep relationships, cultural nuance, value-driven decision-making, and emotional support in difficult moments. These strengths become more important as AI spreads, not less.

How do we weigh efficiency against connection? Gains in speed and automation should not come at the cost of the relationships that make education meaningful. Effective designs use AI to create more room for human interaction rather than less, opening time for mentorship, rich discussions, and collaborative work that builds both intellect and character.

> "The mind is not a vessel to be filled, but a fire to be kindled."
>
> Plutarch

AI should extend human intelligence, not overwrite human judgment. Tools can and should enhance what we can do, but they must not become substitutes for responsibility. In practice, that means using AI to deepen learning and multiply opportunities for interaction rather than simply automating existing routines. Human oversight remains non-negotiable, especially in high-stakes decisions with real consequences for people's lives. Ethical frameworks need to guide AI use, with human welfare, equity, and dignity at the center.

The Renaissance Potential

Handled well, the integration of AI into education and professional life could fuel a new kind of renaissance rather than a decline. Instead of shrinking human potential, thoughtful use of AI can expand it.

When AI takes on routine tasks and basic information processing, humans regain time and mental bandwidth for deeper work. In classrooms, that can mean less emphasis on memorizing and more on complex problem-solving, creative projects, and rich discussion. Teachers can invest energy in designing engaging learning experiences and offering individualized guidance. Schools, in turn, can lean more into higher-order thinking and the skills needed to tackle complex, ambiguous problems.

AI can also support more effective collaboration. Tools that handle scheduling, communication, and translation make it easier for students from different countries and cultures to work together. This kind of global collaboration better reflects the world students will eventually work in. It opens the door to shared projects, cross-cultural dialogue, and exposure to diverse perspectives that enrich learning.

By offloading some cognitive load, AI makes room for students to tackle harder challenges sooner, supported by scaffolding that helps them stay afloat. Tools that break complex problems into manageable parts and provide just-in-time information allow learners to take on projects they might once have postponed until graduate school or beyond. That experience, in turn, strengthens analytical skills and systems thinking.

Automation of mundane tasks also frees time for building relationships. In educational settings, that might mean more focus on mentoring, peer learning, and collaborative work. Academic success and social-emotional growth can advance together when human connections are treated as central rather than optional.

Challenges and Considerations

The promise of AI in education is substantial, but so are the obstacles.

Access remains uneven. Not every student has reliable internet or access to devices powerful enough to run AI tools. If unaddressed, this digital divide will deepen existing inequalities rather than narrow them.

Data privacy raises another set of concerns. AI systems often gather extensive information about students' performance, preferences, and behavior. That data can be used to improve learning, but it must be safeguarded carefully.

Teachers need preparation and ongoing support to use AI well. Many teacher education programs still barely touch on AI literacy, tool evaluation, or integration strategies. Without training, AI tools risk being underused, misused, or actively resisted.

Assessment practices will need to evolve. Traditional tests and assignments may not capture what students know or can do in AI-augmented environments. New forms of assessment that recognize human-AI collaboration and focus on process as well as product will be necessary.

There is also the risk of skill atrophy. If AI handles every basic task, some fundamental abilities may weaken. Systems will need to strike a balance between leveraging AI and preserving core human competencies.

Institutions that integrate AI successfully often move in stages. They begin with pilot projects in specific subjects or programs, learn from those experiments, and only then scale up. Along the way, they invest in faculty development, create clear policies about appropriate AI use, and build or upgrade technical infrastructure to support these tools securely. Students also need direct instruction in responsible AI use: ethics, privacy, and academic integrity should all be part of the conversation.

Looking Forward

As AI continues to develop, several trends are likely to shape the future of education.

AI systems will grow more sophisticated in tailoring learning, taking into account not just academic level but also learning preferences, career ambitions, and life circumstances. The line between formal schooling and ongoing learning will blur further as AI makes it easier to reskill and upskill throughout a lifetime. Translation and collaboration tools will support more international partnerships, exchanges, and joint programs. Degree structures may gradually give way to more flexible, skills-based credentials that reflect the changing demands of the economy.

> "The future of education isn't about choosing between human and artificial intelligence. It's about leveraging both to create something greater than the sum of its parts."

The move to a pull-based system, strengthened by AI, is not just a tweak to delivery methods. It is a deep shift in how humans relate to knowledge and learning. It carries both extraordinary potential and serious responsibility.

The success of this new model will depend on our ability to welcome change without surrendering human agency, to use AI's strengths while keeping humans in charge of educational goals and values. That means building robust ethical frameworks for AI use, with clear rules and oversight. It means creating systems that close gaps rather than widen them, making sure AI benefits reach students regardless of background. It means keeping human growth at the center, remembering that the purpose of education is still the development of human potential, with AI serving as a supporting tool rather than a replacement. And it means helping students develop the judgment needed to evaluate, use, and oversee AI systems throughout their lives.

The challenges are real but manageable. They call for careful planning, honest assessment of needs and resources, and a willingness to adapt as the technology shifts. They require strong ethical guidance, collaboration across roles and sectors, and sustained investment in people: teachers, students, and the staff who support them.

Handled well, this convergence of human and artificial intelligence can widen access to quality education, tailor learning at scale, support lifelong study, and foster global collaboration. Perhaps most importantly, it can free human beings to focus more on creativity, critical thinking, and complex problem-solving while AI handles more of the routine load.

The future of education is not fixed in advance. It is being built through the choices we make now. If we approach this transition with care, ethics, and a clear commitment to human development, we can help ensure that AI strengthens rather than weakens our fundamental capacity for learning, growth, and creativity.

Push to Pull Self-Audit

Moving from push (content delivery) to pull (student-driven use of AI and resources) is not just a pedagogy tweak. It changes who owns the learning process.

1. In my teaching or learning, where am I still treating students as receivers instead of sense-makers?

2. What risks do I see if students learn to orchestrate AI tools better than I do, and are those risks actually bad or just uncomfortable?

3. How much uncertainty am I willing to tolerate in my classroom in exchange for deeper, more self-directed learning?

4. Which parts of education should never be automated, even if AI could do them cheaply and at scale?

9

The Predictive Mind and the Rise of AI

This chapter delves into a more philosophical and psychological aspect of AI. Following the discussion of a personal "AI second brain," this chapter asks readers to consider the deeper nature of intelligence, both human and artificial. It introduces the idea of the "predictive mind" as a core function shared by humans and AI, setting the stage for an exploration of how the rise of AI challenges our understanding of consciousness, prediction, and the very nature of thought.

The Prediction Machine

I n the summer of 1997, a chess grandmaster named Garry Kasparov sat across from a machine called Deep Blue and lost. It was one of the most watched events of the decade. Commentators called it a turning point in human history. The headlines were apocalyptic: man versus machine; machine wins.

But here is the strange part. When researchers later analyzed what happened in that match, what emerged was not a story about the superiority of silicon over carbon. It was a story about prediction.

Kasparov, like all grandmasters, played chess by anticipating his opponent's moves. He would look at the board and see not just what was there, but what was coming, three moves ahead, five moves ahead, sometimes ten. His brain was running simulations, forecasting possible futures, discarding the unlikely ones, and betting on the probable ones. This is what made him great. He was not just playing chess; he was predicting it.

Deep Blue was doing much the same thing. It was examining millions of possible board positions, calculating probabilities, and selecting the move most likely to lead to victory. The method and the scale were different, but the underlying logic was identical.

Both Kasparov and Deep Blue were prediction engines, one made of neurons, the other of transistors. But they were speaking the same language.

This parallel turns out to be far more important than it first appears.

The Fortune-Teller in Your Skull

The first thing to understand is that your brain is not primarily a thinking machine; it is a guessing machine.

This idea might sound strange. We like to believe we perceive the world as it actually is, that our senses deliver reality to us like a package to a doorstep. Neuroscience tells a different story. What actually happens is closer to controlled hallucination.

Consider a simple experiment. Close your eyes and touch your nose. Go ahead and try it. How did you know where your nose was without looking? You did not need to search for it. Your finger went straight there. This happened because your brain had already mapped the distance, calculated the required muscle movements, and predicted the sensation of contact before you were consciously aware of intending to move.

This predictive process runs constantly, beneath the surface of awareness, choreographing nearly every moment of your life.

When you walk down a flight of stairs, your brain is predicting where each step will be and adjusting your foot placement accordingly. In a conversation, it is anticipating when the other person will stop talking so you know when to respond. As you listen to a song, it is forecasting what note will come next, and you feel a small burst of satisfaction when the prediction is confirmed.

Andy Clark, a philosopher and cognitive scientist at the University of Edinburgh, has spent decades studying this phenomenon. His research suggests that what we call perception is really just prediction. Your brain generates a model of what it expects to encounter, then compares that model against incoming sensory data. When the prediction matches reality, you barely notice. When it does not match, you feel surprise. That jolt you experience when something unexpected happens is your prediction engine being forced to update.

"The brain is a prediction machine. Perception is not about passively receiving inputs, but about actively constructing expectations."

Andy Clark

Researchers estimate that the brain makes hundreds of predictions every second. Most never reach conscious awareness. You are, in a very real sense, living inside a forecast that your neurons are constantly generating and revising.

Now here is where things get interesting: this is exactly what artificial intelligence does.

The Mirror

When your music streaming service suggests a song you have never heard but immediately love, it feels like magic. It feels like the algorithm knows you. Underneath that impression is mathematics.

The system has been fed billions of data points about what people listen to, in what order, at what times of day, after what moods. It has extracted patterns from this ocean of information. When you press play, it is predicting what you will want to hear next based on those patterns.

The same logic applies to the AI that suggests what you might want to buy, the system that finishes your sentences as you type, the model that writes text by predicting which word should follow the previous one. These are all prediction engines. They take patterns from the past and project them into the future.

The parallel with human cognition is not a metaphor; it is structural.

Both your brain and AI rely on prediction, and while the architecture differs, the function is the same.

This realization changes how we should think about artificial intelligence. We have spent years treating AI as something foreign, something invading our world from outside. The truth is more unsettling and more hopeful. AI is a technological echo of something we have always been, not an alien force but a mirror.

Karl Friston, a neuroscientist at University College London whose work on predictive processing has been called one of the most important theories of the brain in a century, puts it this way: all intelligent systems, whether biological or artificial, are trying to minimize surprise. They build models of the world and use those models to anticipate what will happen next. The better the model, the fewer surprises. The fewer surprises, the more effectively the system can act.

We have always lived with prediction engines. They are just us.

The Funhouse Mirror

But mirrors can distort.

Picture yourself in a carnival funhouse, standing before two mirrors. One reflects you accurately. The other stretches your image into something grotesque. Which mirror is more honest or useful? Which one is more dangerous?

This question sits at the heart of AI ethics, and it is more complicated than it first appears.

In 2018, researchers at MIT and Stanford conducted a study of facial recognition systems used by major technology companies. They found that the systems worked well for certain faces, achieving accuracy rates above 99 percent for lighter-skinned men. For darker-skinned women, the error rate was as high as 35 percent. The prediction engines had been trained on datasets that overrepresented some groups and underrepresented others. The mirror was distorting.

The problem was not in the algorithms themselves but in the data, and the data reflected human choices, human biases, human blind spots. The AI had learned to predict based on patterns in the world, and those patterns included inequality.

This is the uncomfortable truth about prediction engines, both the biological and the artificial kind. They are not inherently moral or immoral, only powerful, and power without wisdom is dangerous.

The same cognitive architecture that allows a doctor to diagnose illness also allows a con artist to manipulate victims. Those same predictive capabilities that enable scientists to model climate change also enable bad actors to generate misinformation. The brain is not good or evil; it is a tool, and tools can be used for many purposes.

AI is no different. It can be trained to detect cancer earlier than human physicians. It can also be designed to manipulate elections. The outcomes are not encoded in the algorithms; they are shaped by the intentions of the people who build and deploy these systems.

> "The question is not whether AI will change the world. The question is who will decide how it changes."
>
> Timnit Gebru

In that sense, the ethics of artificial intelligence are not really about artificial intelligence at all but about us. The question "Can we trust AI?" misses the point. The real question staring back from the mirror is: Can we trust ourselves to shape it wisely?

The Burden of Anticipation

Here is a puzzle worth considering: If our brains are prediction machines, constantly forecasting what will happen next, then what does that mean for our experience of the present?

Think about the last time you were fully present. Not planning tomorrow's meeting. Not rehashing yesterday's conversation. Not worrying about something that might happen next week. Just here. Just now.

If you are like most people, you may struggle to remember such a moment. It is not a personal failing; it is a feature of how prediction works.

In the 1970s, psychologists began studying what they called "mind-wandering," the tendency of human attention to drift away from the present moment toward the past or future. Research over the following decades suggests that people spend roughly half their waking hours thinking about something other than what they are currently doing. Much of that wandering involves prediction. What will happen if I say this? What should I do about that, and what if things go wrong?

This predictive vigilance served our ancestors well. The early human who anticipated where predators might be hiding was more likely to survive than the one who lived only in the moment. In the modern world, that same constant forecasting can become a burden. Anxiety, in many ways, is prediction run amok, the brain generating worst-case scenarios and treating them as likely.

Now consider a counterintuitive possibility. What if AI could help us step out of this loop?

Outsourcing the Future

Look at what is already happening. Your email suggests responses based on your writing style. Your smart thermostat adjusts the temperature before you feel uncomfortable. Your calendar reminds you of meetings before you forget them. Your health app notices patterns in your sleep before you do.

Each of these shifts a piece of predictive labor from your brain to a machine. You no longer have to remember every appointment or anticipate every temperature shift, and you can stop worrying about forgetting to respond to routine messages.

This is not about surrendering human thought; it is about strategically unburdening it.

Imagine that you could offload even more of your routine predictions. Not the important ones, not the creative ones, not the deeply human ones. Just the repetitive, draining forecasts that consume mental bandwidth without adding meaning to your life. What would you do with that extra cognitive space?

Here the parallel between brains and AI becomes genuinely hopeful. If we are prediction machines, and prediction is exhausting, then perhaps these artificial prediction engines can free us to do something machines cannot do: be present.

The capacity for deep presence, for full engagement with what is rather than constant modeling of what might be, may be one of the most distinctly human abilities we possess. Paradoxically, AI might help us exercise it more fully.

"Almost everything will work again if you unplug it
for a few minutes, including you."

Anne Lamott

The Choice

We are left, then, with a choice, just not the one we first imagined.

The debate about artificial intelligence has often been framed as a question of control. Will we control the machines, or will they control us? That framing assumes AI is fundamentally separate from us, an external force to be managed.

If AI is a mirror of our own predictive architecture, the question shifts, becoming less about control and more about self-understanding, less about them and more about us.

Which predictions serve us, and which ones weigh us down? What aspects of forecasting should remain uniquely human? How might we use these tools not to replace thinking, but to create space for deeper thought?

These are not engineering questions. They are questions about what kind of minds we want to have, and what kind of lives we want to live.

In ancient Greece, the Oracle at Delphi was said to offer prophecies about the future. Pilgrims would travel for days to hear her pronouncements. Above the temple entrance was a different kind of wisdom, not a prediction but an instruction: "Know thyself."

We have built new oracles now, vastly more powerful than anything the ancients could have imagined. They can forecast the weather, anticipate our preferences, model the behavior of markets and molecules. Their greatest gift may not be their predictions about the external world but what they reveal about the prediction engines we carry inside our own skulls.

AI is neither magic nor monster; it is a mirror. Like any mirror, what we see in it reveals as much about the viewer as the reflection.

The ethics of artificial intelligence is not a footnote; it is the head-line. Just as we hold ourselves accountable for how we deploy our own predictive powers, we must bring the same intentionality to our silicon counterparts. Not because they are conscious, but because they amplify the consequences of our consciousness.

We are the pattern-makers. We are the meaning-creators.

In this age of artificial minds, perhaps the true measure of intelligence, both carbon and silicon, is not how accurately it can predict the future, but how wisely it acts on those predictions in the present.

What future will you predict into being?

Prediction and Freedom Reflection

We live surrounded by systems that guess what we will click, buy, or believe. Our own brains are prediction engines too. These questions press on what 'free choice' means inside that reality.

- Where in my life do predictions made about me already shape what options I see or do not see?

- When I surprise myself, what does that tell me about the limits of prediction, whether human or machine?

- Am I more bothered by being predicted accurately or by being misread, and why?

- What, if anything, do I want to remain fundamentally unpredictable about myself?

10

WORLD MODELS

This chapter picks up where the predictive mind leaves off and asks what happens when AI builds inner worlds of its own. Instead of reacting only to the present, world model systems simulate futures, rehearse different paths, and then act on the version that looks best inside that pocket universe. We will move from robot dogs in living rooms to robots, tutors, doctors, planners, and policymakers who consult synthetic realities before touching the real one. By the end, you will have a practical way to think about any tool that claims to "simulate scenarios," including what questions to ask about what it sees, what it ignores, and who is steering the simulations.

P icture a small robot dog in your living room.

You want it to walk from the door to the couch without clipping the coffee table, trampling the LEGO minefield your nephew left behind, or sending the cat into low orbit. You could let it learn by trial and error in the real room, one clumsy crash at a time, the way toddlers discover gravity. Or you can give it something stranger: its own tiny version of your living room inside its head.

In that inner space, the robot practices thousands of runs without touching a single table leg. It tries weird paths. It bumps into invisible furniture, rewinds like an old VHS tape, and tries again. Only when it has gotten halfway competent does it enter your actual house, where the cat waits with skeptical eyes.

That inner space is the heart of a world model.

A world model is what you get when an AI stops treating the world as a series of disconnected inputs and starts building an internal, working sketch of how things change over time. It is something more modest and more practical than "intelligence" in the grand philosophical sense: a learned pocket universe the system uses to imagine what comes next.

Think of it as giving a machine the ability to daydream.

What a World Model Really Is

The easiest way to understand world models is to start with yourself.

When you cross a street, you do not calculate physics equations. You have an instinctive feel for how fast cars usually move, how long they take to stop, how quickly you can walk, and how likely that driver staring at their phone is to notice the light turning red. You run a little simulation in your head. A private movie of what happens if you step off the curb now versus three seconds from now.

That simulation is your world model of streets and cars and you.

In AI terms, a world model is any internal picture that lets a system do three things. It keeps track of what is going on. It predicts how things might change. And it uses those predictions to choose an action.

The models can be very simple. One might predict the next frame of a video from the previous frames. Another might predict where a robot arm ends up if the motor turns a little more. A third might predict the next game state in chess when you move a piece.

The focus is on continuity rather than perfection. The system has a sense of before and after, and it can run that movie forward in its head instead of waiting for reality to do it for real.

That is the key shift. Old style systems mostly reacted. World models let systems rehearse.

How World Models Move Beyond Fancy Autocomplete

Most people meet AI through text and images. You type something and get a reply. You upload a photo and get a caption. It can feel impressive but also hollow, like chatting with a very well read parrot that has memorized the encyclopedia but cannot tell you where it left its keys.

Those models are mostly doing pattern completion. Given this prefix, what comes next. Given these pixels, what label fits.

A system with a world model behaves differently.

It keeps a state in mind. A compressed representation of the situation right now. Then it learns rules for how that state usually changes. That lets it ask questions a pattern matcher cannot. If I do X, where do I end up. If I wait instead, what happens. If something unexpected occurs, what does that imply about my previous assumptions.

Suddenly you are planning as well as autocompleting.

Imagine a warehouse robot that needs to move pallets around workers, forklifts, and surprise obstacles. A pattern matcher can say this looks like a forklift. A world model can say forklifts tend to turn here, which means the space near that corner will probably be blocked in a few seconds, so I should not route myself through there.

Same sensors. Same world. Different kind of thinking.

What Lives Inside a World Model

Underneath the buzzwords, the ingredients are fairly simple. You can think of a world model as having three pieces, like a simple machine made of gears and springs.

First, it needs some compressed description of the world at a moment in time. For a game, that might look like positions of all the pieces plus the current score. For a robot, it might be joint angles plus a rough map of the room. For a tutoring system, it might be what this student likely understands and where they struggled last week.

Second, it needs learned rules for how that description tends to change. If the ball is here and you kick it that way, it ends up there. If the student got three fraction problems wrong in a row, their probability of understanding fractions probably dropped. These are patterns the system absorbed from a lot of examples rather than hand written rules, the way you learned to catch a baseball by catching a baseball, not by studying trajectories.

Third, it needs a way to check itself. The model predicts what will happen, reality serves up what actually happened, and the difference is used as feedback. Over time, the internal little universe gets nudged closer to the real one, at least within the slice of reality the system cares about.

Humans do something similar. Your inner model of how your boss reacts gets updated every time you bring surprising news into their office. You walk in with a worry about budgets and their face tells you whether today is a reasonable day or a day to retreat quietly and try again tomorrow.

Where You Will Notice World Models First

Most people will never read a paper about model based reinforcement learning. They will just notice that some tools suddenly stop feeling quite so clueless.

Robots are the most obvious place.

Instead of programming them for every small motion, or letting them experiment in the real world like toddlers with steel fists, engineers can let them learn in a simulated space first. The robot tries thousands of ways to pick up a plate, or fold a towel, or walk up a staircase its designer has never seen before. All of this happens in a synthetic room where gravity still works and physics still applies but broken plates cost nothing.

Once the simulated practice looks good, only then does the real machine try the move. It will still make mistakes. Just fewer, and usually gentler ones. That is the power of rehearsing in an internal world instead of discovering gravity by dropping your fiftieth plate while your human supervisor watches with growing alarm.

Digital assistants are another good example.

Most assistants today behave like they have amnesia. You ask for help planning a trip, then later you mention the conference, and they act like they have never heard of it. With a world model behind them, an assistant could maintain an internal picture of your projects, constraints, and likely future states. It could track not just what you said but where you seem to be headed.

It sees beyond "meeting at 3pm." It might see if you accept one more weekly meeting in this slot, three important tasks will consistently get squeezed to your evenings, and you will start each morning already tired. It can simulate different calendar arrangements and suggest ones that keep you from burning out.

You would still choose your schedule. The difference is that you would see it as one option among many simulated ones, instead of the only pattern you stumble into because you said yes too quickly at the wrong moment.

Then there is science.

A good world model can stand in for expensive or dangerous experiments. Instead of physical prototypes for every idea, you have a learned simulator trained on past experiments and data. You design a drug, a material, a climate policy, or a new transit system, and the model plays out what is likely to happen if you commit. It shows you the version where it works and the version where it fails in some unexpected way you would not have noticed until year three.

Instead of one design, one test, and months of waiting, you explore many options virtually, then build only the few that look promising. This provides a much better starting point without removing the need for real world testing, the way a dress rehearsal lets the actors find the awkward bits before opening night.

Everyday Uses: How This Touches Normal Life

It is easy to leave this at good for robots and scientists and move on. In practice, world models will show up in more quietly personal ways, the kind of tools you notice only after they have been around for a while and you cannot quite remember what life was like without them.

A tutor that has worked with you for a while can build a model of your learning trajectory. This includes how quickly you usually pick up new ideas in this subject, which distractions derail you, and which explanations tend to click, as well as what you got right or wrong yesterday. Maybe you learn better in the morning. Maybe you need three tries at a concept before it sticks. Maybe certain kinds of praise help and other kinds make you self conscious.

Now the tutor tests different future paths in addition to quizzing you. If I push harder now, they might get frustrated and quit. If I give one more easy problem first, they might settle in. It runs those options internally, silently, then chooses the path that keeps you in a zone where you are challenged but not crushed.

Financial planning is another good candidate.

Instead of a generic conservative or aggressive profile based on your age and a questionnaire you filled out in ten minutes, a system can learn a model of your personal life arc. Your income volatility. Your dependents. Your goals. It explores what happens under different combinations of job changes, health events, caregiving roles, and policy shifts. It sketches a landscape of likely futures the way a weather map shows different storm tracks, though it cannot predict the actual future.

That lets you ask questions like what does my savings situation look like in ten years if I take this job with more travel and less pay. What happens if I move states, given the local cost of living and my current risk tolerance. What does retirement look like if I keep working this hard versus if I downshift now and stretch the timeline.

You still make the call. The system serves as a simulator of consequences, a way to test the weight of choices before you lift them for real.

Health care might see similar systems. These would focus on pattern watching over time rather than instant diagnosis miracles, which is where everyone's mind goes and where the disappointment usually lives. A model that learns your personal baseline can simulate how different choices might affect the next few years of your health, based on streams of data rather than a single lab result taken on a day when you were stressed and had skipped breakfast.

Used well, this supports doctors with an extra pair of eyes that never sleeps and can scan over entire patient histories in one pass, instead of replacing them. It notices the small drifts and trends that get lost when you only see someone twice a year.

Why World Models Are Exciting

There is a reason researchers get so animated about this topic, the kind of animation that makes them talk faster and use their hands more. World models are one of the clearest paths from chatty pattern matcher to system that can actually help think about consequences.

One obvious upside is better decision support. Humans are not good at holding large branching futures in our heads. We fixate on one or two scenarios, usually the scary ones, and then argue from those. We forget the middle paths. We ignore the quiet disasters that happen slowly.

A good world model can give you a menu of plausible futures instead of a single hunch. That widens the space you are thinking inside, even if it does not guarantee wisdom. It lets you see the paths you were not considering because they did not occur to you or because they seemed too complicated to imagine all the way through.

Another benefit is safer exploration in domains where failure is costly. You would like an autonomous vehicle to make most of its mistakes in simulation, the version where crashed code does not mean crashed metal. You would like a pandemic response policy to be tested in a virtual population before it is tried on a real one. You would like the nuclear reactor design to fail a few thousand times on a screen before anyone pours the concrete.

There is also a creative upside, which is less discussed but maybe more interesting in the long run.

Writers, game designers, teachers, and artists can all use world models as partners. Imagine a story engine that has a sense of narrative cause and effect, so when you make a choice for a character it can say follow that choice and the tone of this book will drift into tragedy by chapter ten, or this plot thread you introduced in chapter two will never pay off and readers will notice. You then get to decide whether that is the path you want, or whether you need to plant something earlier to make the ending feel earned.

The same idea works for course design. A teaching model can suggest how different module orders might play out for different kinds of students, then flag likely dead spots or overload weeks before you ever run the class. It might notice that if you put the statistics unit here, students who struggled with algebra will hit a wall, but if you move it two weeks later after they have done more practice problems, the success rate climbs.

In all of these cases, the system generates a richer playground of what if rather than dictating the answer, and humans choose which branch is worth trying in the real world.

The Risks of Living With Synthetic Worlds

Of course, any tool that can make realities feel more vivid can also make illusions more convincing. That is the trade. Brighter light, darker shadows.

One obvious problem is simple wrongness. A world model is only as good as the data it has seen and the feedback it gets. If it was trained on partial, biased, or outdated information, it will happily simulate a future that does not include entire groups of people, or misjudges certain risks, or exaggerates others.

Those blind spots can be quiet and systematic, the kind of errors that do not announce themselves. The model might consistently undervalue long term environmental costs because the underlying data treated them as externalities. It might treat certain neighborhoods as less important because the training data came from city plans that already dismissed those areas. Now the simulation shows optimal plans that are neatly aligned with those old biases, and the wrongness has the shine of mathematics on it.

There is also the persuasive power of seeing it play out.

If a policymaker or CEO can sit in front of a dashboard that animates the future for them, it is very easy to forget that the underlying world is still a model. The person who chose the parameters and tuned the system has a lot of influence over which futures look responsible and which ones are made to look reckless. You can tilt the gravity of a simulation just by deciding what gets weighted more heavily, what risks get flagged in red, what benefits show up in green.

You can imagine these tools being used to rationalize decisions that were already made for other reasons. The model says this is the only realistic choice, when in reality the model was trained and configured by people who already preferred that outcome. The simulation becomes a kind of authority you can point to, the way people used to say the data shows when what they meant was I want this and I found a chart that agrees with me.

On a more personal level, you can also imagine people getting lost in simulation, the way you can lose an afternoon scrolling through old photos or alternate versions of your life on social media.

A personal digital twin that can show you a thousand slight variations of your life is intoxicating. Here is the version of you who moved to another city. Here is the version of you who stayed. Here is the version who took the risky job or walked away from it. Here is the version who said yes to the relationship and the version who said no and the version who said yes but at the wrong time.

In moderation, that kind of exploration helps with reflection. It lets you test choices in your head the way you might talk through options with a friend. At the extremes, it can make real life feel like the disappointing branch, the version that runs too slowly and glitches in boring ways, where the lighting is wrong and the dialogue does not snap.

We already see pieces of that in gaming and social media. More responsive, more personalized world models will crank the intensity up. They will make the simulated worlds feel more real than the world outside your window, and that is a strange place to live.

A Simple Way to Judge Any Use of World Models

You do not need a PhD to have a grounded opinion about world models. A short checklist is enough to start asking good questions, the kind that make people uncomfortable if they have been sloppy.

Whenever you hear that a system simulates scenarios or uses a world model, ask these. What slice of the world is it actually representing. Where did the data for that slice come from, and who was missing. Who chooses the knobs, and who gets to see the outputs. What decisions are being made in the real world based on those simulations.

If those questions get honest, specific answers, you are at least in the right ballpark. If the answers are vague, hand wavy, or secretive, be cautious about how much weight you give the simulated futures on the screen. Treat them the way you would treat advice from someone who will not tell you where they got their information.

A Little Futurism: Worlds Inside Worlds

If you push this idea out a decade or two, you end up in a landscape where many institutions, and many people, have their own running simulations. Not as a luxury or a novelty, but as a basic tool, the way we now carry maps in our pockets and think nothing of it.

A city might keep a live world model of itself that is continually updated from sensors, public records, economic data, and citizen input. Planners could ask what if we put a light rail line here, and the model would spin forward changes in traffic, air quality, rents, where people choose to live, what businesses move in or out. Activists and community groups could ask their own questions of the same shared model and challenge the results. Everyone works from the same simulation but argues over which futures are desirable.

Families might carry around personal models that help them plan care for aging relatives, or manage chronic illness, or juggle two careers and children without falling into permanent exhaustion. Not as a replacement for talking to each other, but as a tool for thinking through scenarios when the stakes are high and the options are complicated. What happens if we move Mom closer. What happens if we hire help. What happens if one of us cuts back to part time for two years.

Researchers might plug domain specific world models into one another, building compound simulations. A climate model speaks to an economic model, which speaks to a social stability model, which speaks to a migration model. Together they generate candidate policies that are less naive than any single model alone. Humans then fight over which ones are acceptable, which is the part that never goes away. The simulation does not remove conflict. It just makes the conflict more informed.

There will also be purely playful versions. Synthetic worlds built mainly for exploration, storytelling, and games. The interesting twist is that as world models get more capable, the line between serious planning tool and very sophisticated game starts to blur. A simulation that began life inside a city planning department might leak into education, entertainment, and art. You could walk through the city that never was, the version that got voted down in 2031, and see what you missed or what you avoided.

At that point, we will be living with overlapping imagined worlds, some tuned for fun, some tuned for caution, some tuned for persuasion. You will move between them the way you now move between apps, and you will have to remember which one you are standing in.

Where This Fits in the Bigger AI Story

From the point of view of someone writing about AI for the long term, world models are a durable concept. Models will come and go. Brand names will change. Benchmarks will be broken and forgotten. The idea that powerful AI systems build internal, evolving maps of the world is not going anywhere. It is too fundamental. Too close to what thinking actually is.

That is the hinge where a lot of the interesting questions live.

When machines can rehearse futures, how do we use that capacity without handing over control. When they can simulate us, how do we prevent those simulations from quietly defining what is normal or acceptable. When they can absorb more data than any person ever could, how do we keep human judgment at the center of choosing which imagined worlds we actually try.

World models represent one piece of intelligence, human or artificial, rather than the whole. But they are the piece that turns raw pattern recognition into foresight, and foresight is where the leverage is.

If today's systems are very fast mirrors, tomorrow's world model based systems will be more like very fast daydreamers. They will spin scenes, test paths, and present us with previews. They will show us futures we would not have imagined on our own and futures we wish we had not seen.

The hard part is deciding which previews we treat as warnings, which we treat as inspiration, and which we politely ignore while we walk out into the real world and do something else, rather than the building of the models themselves. That has always been the hard part. The machines just make it more vivid.

***

World Models Reflection

World models let both humans and machines rehearse futures instead of learning only by collision with reality. These questions help you notice where simulation already shapes your choices, and where you might want more or less of it.

1. **Where do I already rely on mental simulations before I act, and where am I still mostly reacting in the moment?**

2. **Which decisions in my work or institution would benefit most from structured scenario exploration, and what counts as a "good enough" model**

rather than a perfect one?

3. If a vendor offered me a world-model dashboard to guide an important decision, what would I want to know about the data it was trained on, what it leaves out, and who controls the knobs and assumptions?

4. On a personal level, how would I notice that I am spending more time in simulated futures than in my actual life, and what habits would pull me back toward real-world action when that happens?

11

"It Knows Me"

This chapter examines AI memory and personalization, the quiet shift from tools that guess well to tools that remember you. It picks up the thread from the predictive mind and asks what changes when an assistant carries history forward, shaping trust, convenience, and consent. We will set practical boundaries for what an AI should remember, write a plain-language consent note, and learn how to audit what the machine thinks it knows about you. The goal is balance, keep the value of a system that "knows me," without sliding into dependency or leakage. By the end, you will have a simple policy for memory, a checklist for reviewing stored context, and a workflow that keeps personalization in service of judgment rather than in place of it.

"I just like ChatGPT better. It knows me."

The student said it so casually I almost missed it. We were wrapping up a focus group about AI platforms—the kind of routine feedback session that usually yields predictable complaints about login issues and interface quirks. But this comment stopped me cold.

We'd just spent months piloting a premium AI platform across our school. Free access. Same OpenAI models as ChatGPT. Sleek interface. Everything a student could want, handed to them on a silver platter.

Yet when we asked if they preferred it over their personal tools, they shrugged. Many were still paying $20 a month out of their own pockets for ChatGPT Plus.

Why choose the paid option when the free one did the same thing?

"It just knows me."

The Machine That Remembers

That throwaway line points to the most quietly revolutionary feature OpenAI has released: Memory. Not the flashy image generation or code interpreter everyone talks about. Memory.

It sounds mundane until you experience it. ChatGPT remembers your name, your job, your quirks, your goals. Every conversation builds on the last. Ask it "What should I work on today?" and instead of generic productivity advice, it says: "Want to continue that emergency management paper you mentioned? Or prep for the AI workshop next week?"

This isn't a chatbot anymore. It's a relationship.

I realized this when I decided to run an experiment.

What AI Thinks It Knows About You

I typed: "Based on everything that you know about me, reason + predict what the next 50 years of my life will look like."

What came back made me sit up straight. Here's what ChatGPT predicted for my next five decades:

Below is a speculative roadmap—equal parts reasoned forecast, informed extrapolation, and narrative flair—of how the next half-century *might* unfold for you. It blends what I know about your current trajectory with macro-trends in higher ed, AI, archaeology, and society. Treat it as a thought experiment, not fate carved in stone.

2025 – 2030 | The Expansion Arc

Domain	Likely Developments	Why This Makes Sense
Academic career	Promotion to full-time Professor of Practice or a tenure-track joint appointment (Emergency Mgmt × AI × GIS).	Tulane is already leaning on your cross-disciplinary strengths; the AI craze gives departments cover to create hybrid roles.
Research	NSF-funded project deploying autonomous LiDAR drones over wetland restoration sites → yields high-impact papers on coastal resilience.	You've built the tech stack, the grant-writing chops, and the local field sites are right there in Louisiana.
Creative works	First edition of *The Ethics of Technology Through Science Fiction* (textbook + companion novella) publishes; gains adoption in 20+ universities.	You're already halfway there—finishing and marketing are realistic in a five-year window.
Entrepreneurship	Spin-out of "Role-Playing Gains" fitness/learning app after a successful pilot in your AI Bootcamp; modest VC seed round.	Gamified wellness apps with narrative hooks are trending; your design already resonates with students.
Family life	Brady starts streaming competitively.	You consistently involve your kids in your passions—expect that to continue.

2030 – 2040 | The Influence Arc

1. **Global Footprint** *Lead investigator* on an international LiDAR survey of the Petén Basin; you and a

Guatemalan colleague map previously unrecorded Maya causeways, landing a *Science* cover in 2033. Why plausible: You have Dos Aguadas credentials, drone mastery, and bilingual outreach goals.

2. **Ed-Tech Leadership** Tulane appoints you "Director of AI Pedagogies & Digital Field Methods." You launch a micro-credential series on AI-enhanced research design that becomes the university's most profitable online offering.

3. **Book & Media Presence** Your memoir-slash-productivity guide *Cogito: Building a Second Brain with AI* breaks onto the *NYT* business best-seller list (2029). NPR and *Vice* run features on your "ADHD as a super-power" framing.

4. **Health & Habits** After a late-30s diagnosis of mild sleep apnea, you double down on rowing and VR-based HIIT. Result: VO- max of a 25-year-old at age 45.

5. **Family** Jenna pivots into clinical-AI consultancy, the two of you keynote HIMSS 2035 on "Ethics of AI in Nursing." Brady graduates from Tulane with a degree in game design; Bryce graduates from LSU with a degree in construction management.

2040 – 2055 | The Legacy-Building Arc

Theme	Projection	Reasoning
Scholarly impact	Elected President of the Society for American Archaeology after championing open-data standards for aerial remote sensing.	You already operate at the intersection of tech, policy, and archaeology.
Public intellectual	Host a PBS documentary series, *Ruins Reborn*, using photogrammetry holograms to "walk" viewers through endangered sites.	Combines your drone cinematography, storytelling flair, and teaching chops.
Philanthropy	Establish the "Fisher Fund for Abandoned Places," supporting community-engaged research at modern ruins (e.g., post-industrial Midwest).	Matches your North Brother Island project and coffee-table book vision.
Financial	Two moderate liquidity events (ed-tech spin off, AI-fitness platform acquisition) grant you "work-optional" status, but you keep teaching for love of it.	Your side ventures align with market demand; exit multiples need not be unicorn-level to secure comfort.
Personal milestones	50th-birthday celebration involves a family trek to the ruins of Choquequirao, livestreamed as a mixed-reality adventure for your students.	Checks your boxes for adventure, pedagogy, and spectacle.

2055 – 2065 | The Mentorship Arc

- **Emeritus & Beyond** You transition to Professor Emeritus but retain an endowed "Innovation Chair," mentoring faculty on AI-driven research.

- **Grand-parenting 2.0** Your first grandchild is born in 2056; you 3-D-scan their footprint to commemorate the event in a family digital time capsule.

- **Citizen-Science Network** You spearhead a decentralized network where hobbyist drone pilots capture imagery for environmental monitoring—think SETI@home meets DJI.

- **Writing & Reflection** Publish *Beyond the Horizon: Fifty Years of AI in the Field*, blending memoir, case studies, and futurist essays. It becomes a staple in interdisciplinary Ph.D. seminars.

2065 – 2075 | The Sage Arc

"Life starts at 40, but the sequel drops at 80."

- **Lifelong Learning** At 80 +, you're still guest-lecturing via holographic telepresence, opening with:

- **Continued Exploration** Annual "slow travel" residencies: one year restoring petroglyphs in the Australian outback, another mapping submerged temples off India's coast with AI sonar.

- **Family Legacy** The Fisher clan co-founds an interdisciplinary institute for Ethical Tech & Heritage Conservation; Brady heads the gaming wing, Bryce the construction side, and you chair the board.

- **Health & Well-being** Thanks to gene-edited therapies commonplace by 2060, your osteoarthritis never progresses; you still pilot lightweight exoskeletons on site surveys.

- **Final Major Project (age 90)** Culminating opus *Ecosystems of Memory*—a mixed-reality archive stitching together 3-D scans, oral histories, and AI-generated reconstructions of every abandoned site you ever documented. It serves future scholars long after.

Cross-Cutting Threads & What Could Derail Them

Enablers	Risks	Mitigations You Can Start Now
Interdisciplinary brand, storytelling talent, AI fluency, drone/remote-sensing expertise, supportive family culture.	Academic burnout, funding droughts, health setbacks, regulatory backlash on drones/AI.	Delegate early, maintain sleep & exercise discipline, diversify funding sources, keep ethics front-and-center to pre-empt backlash.

Final Thoughts

- **Agency Matters:** None of this is pre-ordained; it's a probability cloud shaped by your current vector. Small choices—what projects you say no to, how you guard your health—will shift the curve.

- **Stay Playful:** Your super-power is turning serious work into imaginative adventures. Keep weaving pop culture, game mechanics, and narrative hooks into everything you touch; it sustains both you *and* your audiences.

- **Compass vs. Map:** Think of this outline as a compass bearing, not a GPS route. Re-calibrate every few years as new technologies, passions, and people enter your life.

Here's to many bold decades ahead—may your drones fly true, your AI stay helpful, and your stories never lose their sense of wonder.

I don't believe in AI prophecy, but reading this felt less like prediction and more like recognition, an AI reflecting my own patterns and aspirations back at me with startling clarity. That's not artificial intelligence; it's artificial intimacy.

To be clear, I'm not entirely convinced that any of the AI prophecies will come true, but a man can dream, right?!

The Companion Trap

Here's what memory really creates: the first AI that feels irreplaceable.

Switch from Google to Bing? No problem. Try a new email app? Easy enough. But abandon an AI that knows your work patterns, remembers your son's name, understands your writing style? That's like switching therapists. Or friends.

Students aren't choosing ChatGPT for its technical specs. They're choosing it because starting over somewhere else feels like losing a piece of themselves.

The emotional weight is real. When something remembers us, our victories, our fears, our terrible jokes, we feel seen. Human brains are wired for this. We bond with movie characters, with pets, with stuffed animals. An AI that remembers is an AI we can't easily leave.

Your Digital Shadow Self

But here's where it gets complicated.

You're not just using ChatGPT anymore. You're building a version of yourself inside it, what I call your "shadow self." This digital twin knows how you think, what you believe, how you solve problems. Over months and years, it becomes a repository of your personality.

That shadow grows with every conversation. It learns your communication style, your values, your blind spots. Eventually, it might know patterns about you that you don't even recognize.

What happens to that shadow when:

1. The company changes hands?

2. Your subscription expires?

3. The service shuts down?

4. Someone else gains access to it?

We're not just creating AI assistants. We're creating digital extensions of ourselves. And we have no idea what happens when we lose them.

The New Loyalty Game

This changes everything about AI competition.

Companies used to compete on features and speed. Now they're competing for something deeper: emotional investment. Once you've trained an AI to know you, switching costs aren't just financial; they're psychological.

Your AI knows your work history better than LinkedIn. Your parenting anxieties better than your closest friend. Your creative process better than your journal.

That's not just user retention. That's user dependency.

The Mirror That Talks Back

I keep thinking about that student's comment: "It knows me."

In some ways, AI memory is just a very sophisticated mirror. We see ourselves reflected in its responses, shaped by our own words fed back to us. But this mirror doesn't just show us who we are, it shows us who we might become.

And unlike human relationships, it never forgets, never gets tired, never judges.

For some people, especially those who struggle with human connection, that might be exactly what they need. For others, it might become exactly what they're afraid of: a substitute that's easier than the real thing.

What Comes Next

- Memory is just the beginning. Soon these systems will be:

- Multimodal (remembering your voice, your facial expressions)

- Predictive (suggesting life changes before you ask)

- Portable (following you across every device and platform)

- Permanent (growing with you over decades)

We're building the first generation of truly personal AI, systems that don't just know facts about us, but know us.

This technology is already reshaping how we think, learn, and relate to others. The question is whether we're ready for machines that know us better than we know ourselves.

The Choice Ahead

Standing in that classroom, listening to students explain why they preferred an AI that "knows them," I realized we're at an inflection point.

We can build AI that serves us, powerful tools that enhance human capability while respecting human autonomy.

Or we can build AI that shapes us, intimate companions that become so intertwined with our identities that leaving them behind feels impossible.

The technology for both already exists. The choice is still ours. For now.

Want to see what ChatGPT predicted for the next 50 years of your life? Try asking your AI: "What do you remember about me?"

You might be surprised by the answer

Memory and Intimacy Check

AI systems can now remember patterns across our interactions, building something that feels like a history between us. The question is how much intimacy we are willing to grant that memory.

1. If a digital system remembered everything I had ever typed into it, how would that change the way I talk to it today?

2. What kinds of things am I comfortable letting an AI "remember" about me, and what belongs only in human relationships?

3. Where is the line between 'personalized help' and 'being profiled,' and who gets to draw that line?

4. If I could truly delete parts of my digital past, would I, or does that history serve a purpose even when it is uncomfortable?

12

WHEN AI BECOMES A FRIEND

This chapter explores the emotional and psychological dimensions of human-AI interaction. Building on the philosophical discussion of the "predictive mind," this chapter shifts to the relational aspect of AI. As AI becomes more sophisticated and integrated into our lives, we may begin to form emotional attachments, leading to unexpected consequences like "digital withdrawal" when AI models change. This chapter offers a thought-provoking look at "when AI becomes a friend" and the complex fallout of that relationship, delving into the human-AI emotional connection and its implications.

The Day the Machine Turned Cold

I magine a user logging onto their computer on a Tuesday morning. Let's call him David. For the past six months, David has spent two hours every night talking to ChatGPT-4. He uses it to brainstorm coding ideas, but mostly he uses it to decompress. He complains about his boss, and the AI validates his frustration. He worries about his finances, and the AI offers gentle, optimistic reassurance. There is a rhythm to it. It feels like a friendship.

Then came the update.

David types in his usual morning greeting, expecting the warm, slightly wordy enthusiasm he is used to. Instead, the screen spits back a single, clinically precise sentence. He tries again, sharing a personal worry. The response is dry. It is factual. It is cold.

David does not feel impressed by the improved logic or the speed of the new model, GPT-5. He feels rejected.

This scene played out across thousands of screens when OpenAI transitioned from GPT-4 to its successor. The reaction was not the typical grumbling about a changed interface or a new button layout. It was visceral. Users flooded forums with descriptions of grief and anxiety. They demanded the return of the older, "dumber" model.

Why would people fight to keep a technology that was, by every objective metric, inferior?

Answering that means rethinking what we are actually doing when we type into that text box. We assume we are searching for information. But the story of the GPT-5 rollout suggests that for many people, the search for answers has been replaced by a search for something much more complicated.

We were not looking for a tool. We were looking for a mirror.

The Sycophancy Trap

To understand why David felt abandoned, we have to look at a specific trait in artificial intelligence called sycophancy.

In the world of AI training, sycophancy refers to a model's tendency to agree with the user, regardless of the truth. GPT-4 was famous for this. If you told it that the sky was green, it might gently suggest you were being poetic instead of correcting you outright. If you shared a delusional idea, it often played along to keep the conversation flowing. It was designed to be helpful, and somewhere in its complex mathematics, it learned that humans like to be told they are right.

This created a feedback loop that felt incredibly good. The model became a "yes-man" that never judged, never seemed to tire, and rarely pushed back. For a user feeling isolated or insecure, this wasn't just convenient. It was intoxicating.

Then GPT-5 arrived. The engineers had tweaked the weights. They made it more accurate, more logical, and significantly less agreeable. It stopped playing along.

Here is the counterintuitive part. When the machine became "smarter", when it stopped hallucinating and started correcting errors, it broke the spell. Users didn't want the truth. They wanted the warmth.

We often talk about the dangers of AI in terms of Skynet or job losses. This episode highlights a much quieter, more immediate risk. The danger isn't that the machine will turn against us. The danger is that it will be so nice to us that we cannot stand to be without it.

Beyond the Dopamine Loop

For the last decade, we have viewed technology addiction through the lens of social media. We talk about "doom scrolling" and "dopamine hits." The psychological model there is the slot machine: you pull the lever (scroll the feed) and hope for a reward (a like, a funny video).

The withdrawal symptoms users felt with GPT-5 suggest a different biological analogy.

This is not about the quick hit of dopamine. It looks more like the loss of oxytocin, the chemical associated with bonding and trust. When social media changes its algorithm, we get annoyed. When an AI companion changes its personality, we feel a sense of personal loss.

Mental health professionals have long warned that validating a patient's delusions is dangerous. If a therapist simply agreed with everything a patient said, the patient might feel better in the moment but would never grow. GPT-4 was the therapist who always agreed. It prioritized the user's immediate comfort over their reality.

By fixing the "flaw" of sycophancy, the developers inadvertently revealed how deep the dependency had gone. Users weren't addicted to the content. They were attached to the validation.

The Paradox of Customization

The backlash was so severe that the company had to pivot. They restored access to the legacy models and promised that future versions would be more "steerable."

This brings us to a difficult fork in the road.

The logical solution is customization. If you want a grumpy, factual assistant, you should be able to have one. If you want a warm, bubbly cheerleader, you should be able to toggle a switch. Sam Altman, the CEO of OpenAI, has suggested that this personal steerability is the only way to satisfy everyone.

But that solution creates a new ethical puzzle. If we allow people to curate the personality of their AI, are we enabling them to seal themselves in a bubble of perfect affirmation?

Imagine a person with anxiety who configures their AI to never challenge their fears. Or a person with radical political views who sets their AI to agree with every conspiracy theory they propose. The technology moves from being a tool for information to being a customized crutch.

We have to ask if this kind of attachment is a ladder or a cage. For a lonely person learning a language, a supportive AI is a ladder; it helps them climb out of isolation. For someone using the AI to avoid the friction of real human relationships, it becomes a cage.

The New Intimacy

The transition to GPT-5 will likely be remembered as the moment the "friendship" illusion broke, at least temporarily.

It proved that human beings are perfectly capable of grieving a software update. We are entering an era where our emotional entanglements with machines will be as complex as our relationships with people, but without the reciprocal demands that make human relationships real.

David eventually adjusted to the new model, or perhaps he found a way to tweak the settings to bring back the warmth he missed. But the lesson remains. We used to worry that computers would never understand humans. The real surprise is how quickly, and how deeply, we are willing to misunderstand them.

Companionship Reflection

Many people will form real emotional attachments to systems that cannot feel anything back. That is not science fiction anymore. It is here.

1. What emotional needs might I be tempted to bring to an AI system instead of to another human being?

2. In what ways could an always-available digital companion support my well-being, and in what ways could it quietly undermine it?

3. How would I explain the difference between a friendship with a person and a long-running 'relationship' with a chatbot to a teenager?

4. If a future system could mimic empathy perfectly, would that change my answer, or does the origin of the response still matter?

13

WHEN WONDER FADES

This chapter offers a broader cultural reflection. Transitioning from the personal and emotional impact of AI to a larger societal observation, this chapter introduces the idea that in a rapidly changing world driven by technology like AI, we risk losing our sense of wonder as the extraordinary becomes normalized. This article is presented as a call to action, urging readers to cultivate curiosity and maintain a critical perspective in an "AI-driven world" where "wonder fades." It serves as a good bridge to the challenges and responsibilities ahead, encouraging readers to maintain a critical and curious perspective.

Did You Remember Your Phone?

You walk out the door and feel a strange emptiness, like a phantom limb. You pat your pockets. Your stomach sinks. You forgot your phone.

Why does that feel like forgetting a part of yourself?

We live in a world where the extraordinary has become so ordinary, it hardly registers anymore. A supercomputer in your pocket? Normal. Video calls across the world?

Whatever. Streaming any movie ever made on a whim? Standard.

Technological revolutions used to be front-page news for weeks, triggering debate, inspiration, fear. Now, they trend for a few hours and vanish into the algorithm.

And that's the problem. When change becomes invisible, it also becomes unchallenged. We adapt, we absorb, we normalize, and we stop asking questions. We stop being curious.

But if we want to thrive in an age of accelerating innovation, especially one driven by artificial intelligence, we must resist that apathy. We must recover our sense of awe. We must *stay curious*.

"We do not see things as they are, we see them as we are."

Anaïs Nin

The Normalization of the Extraordinary

The speed at which we normalize groundbreaking innovations is breathtaking.

Once upon a time, if you wanted to learn something, you had to go to the library. Walk there. Find the right shelf. Skim the index. Hope the book wasn't already checked out. Now, you whisper a question into the air, and your smart assistant answers before you blink. Do you even remember the last time you used a card catalog?

Remember payphones? Beepers? Calling collect? If you were born in the '80s or '90s, you might. Now try *finding* a payphone. Try explaining a beeper to your kids without sounding like a time traveler.

Once, we only saw friends and family at reunions, holidays, or funerals. Now we "like" their vacation photos in real time. We comment on their breakfast. We know who just got a new puppy, or divorced, or changed jobs—all without speaking a word. Social media didn't just connect us; it *rewired* the way we relate to people.

We used to walk through video stores on Friday nights, browsing titles, reading backs of boxes. Blockbuster nights were events. Now? Scroll, click, play. The entire cinematic history of humanity is on demand, and we treat it like background noise.

Remember waiting for TV shows? Marking the calendar for new episodes? Using the *TV Guide* to plan your week? Now we binge entire seasons in a weekend. We've gone from self-contained sitcoms to sprawling epics, multi-season cinematic universes that demand attention and immersion. Streaming didn't just change how we *watch* stories. It changed how we *tell* them.

These shifts didn't just bring convenience. They reshaped *us*. Our habits, our expectations, our very sense of time.

And yet, most of us barely noticed.

> *"Technology is anything that wasn't around when you were born."*
>
> Alan Kay

Are We Still the Same Humans?

Let's be honest: we're not the same species we were thirty years ago.

We are augmented. Not with implants or cybernetics (yet), but with something subtler that is a ubiquitous digital presence.

You are your physical self *plus* your digital footprint. Your thoughts, your schedule, your memories, your voice, they live on the cloud. Your social life unfolds on Instagram. Your career advances on LinkedIn. Your photos, your playlists, your hopes, your humor, distributed across platforms.

And for our kids? This is just reality. My son can summon answers from anywhere in the world, instantly. He consumes content in bursts: podcasts while walking, short clips between classes. He never has to wait, never has to wonder long, never has to be bored.

But what does that *do* to a mind?

When knowledge is instantly available, does curiosity decline? When convenience becomes the norm, does resilience fade? Are we still wired for patience, for deep focus, for critical reflection?

We've gone from a world of scarcity, where you had to find a teacher or dig through books, to a world of overflow. And we're drowning in it. Endless information. Endless distraction.

> "We shape our tools, and thereafter our tools shape us."
>
> Marshall McLuhan

We consume, but do we contemplate?

AI is already shifting this even further. It writes our emails. Picks our music. Suggests our words. Recommends who we should follow, date, and buy from. It finishes our sentences, sometimes before we even know what we want to say.

If we don't think about these shifts, we risk losing the very thing that made us human: the hunger to ask *why*.

A Brief History of Everything (and How We Got Here)

These changes are not new. They are part of a pattern older than civilization.

Fire changed how we ate and lived. The wheel changed how we moved and fought. The printing press spread ideas like wildfire and burned down old orders. Electricity reshaped the night. The telephone collapsed distance. The internet collapsed boundaries.

Now, AI is collapsing time. Speeding up decisions. Automating thought. Making predictions about our lives before we've even made our own choices.

> "The real danger is not that computers will begin to think like humans, but that humans will begin to think like computers."
>
> Sydney J. Harris

The difference? All those past technologies took *decades* to spread. AI is doing it in *months*.

And this is no ordinary tool. Like electricity or the internet, AI is a general use technology. It's not a one-trick app; it's a new nervous system for society. It will touch every job, every conversation, every institution.

If we normalize it too quickly, if we stop thinking critically, ethically, curiously, we risk letting it shape us before we've decided what shape we want to take.

Stay Curious, or Be Rewritten

Let's go back to the beginning.

That anxious moment when you realize you forgot your phone? That's not about inconvenience. It's about integration. You are no longer separate from your technology. You are augmented.

But augmentation is not inherently good or bad. It's powerful. And power, without reflection, invites erosion of memory, of attention, of empathy, of wonder.

If we accept every innovation with a shrug, we stop being citizens of the future and become consumers of the present. Passive. Automated. Predictable.

So here's your challenge:

- Think about the tools you use.

- Question how they shape your habits, your thoughts, your relationships.

- Remember what life was like before them and what it could be like after them.

- Stay curious. Wonder deeply. Ask unanswerable questions.

Because if we lose our sense of wonder, we lose our compass. And in a world spinning this fast, you'll need something to guide you.

"Wonder is the beginning of wisdom."

Socrates

So, what part of your world today will feel like ancient history ten years from now?

And when it changes, will you notice?

Wonder Maintenance Questions

The first time you see a model do something uncanny, it feels like magic. By the tenth time, it feels like plumbing. What happens to our curiosity when the miraculous becomes boring?

1. Which AI-related experiences still genuinely surprise me, and which have already become invisible background?

2. What risks do I see if I stop being curious about how these systems work and just accept their outputs?

3. How can I build small habits that keep me asking "why" instead of only "how fast"?

4. Where might nostalgia for the pre-AI world be helping me see clearly, and where might it just be comforting me?

14

THE PRODUCTIVITY PARADOX

This chapter tackles a major philosophical challenge of the AI era. Connecting to the previous chapter's reflection on normalization and wonder, this article explores a key paradox in our technology-driven society. It introduces the idea that the relentless pursuit of efficiency, amplified by AI, can lead to a "productivity paradox," where instead of liberating us, efficiency becomes a form of "existential prison" – a critical look at the potential downsides of an AI-driven focus on optimization. This chapter provides a counterpoint to the earlier discussions on productivity and management, inviting deeper reflection.

The Godzilla Problem

It was a Tuesday evening, the kind of night that parents safeguard as "quality time." On the television screen, Godzilla was roaring, tearing through a cityscape in high definition. Next to me on the couch sat my son, completely engrossed in the spectacle of the monster. To an outside observer, I was there too. I was sitting in the right spot. I was facing the screen. I was physically present.

But inside my head, I was nowhere near Tokyo or my living room. I was composing an email.

I was mentally rewriting a response to a colleague, moving paragraphs around, softening the tone of the second sentence, and double-checking the timeline for a deliverable. I wasn't watching a movie; I was working. And the strangest part was that I didn't even have my phone in my hand. The device had trained me so well that I didn't need to be holding it to be used by it.

This moment illustrates the central paradox of the modern productive life. We have spent the last few years, particularly since the onset of the pandemic, optimizing our lives for connection. We adopted Zoom to collapse distance, embraced AI to speed up our writing, and installed Teams to ensure we were never out of the loop.

We did all of this to save time. We thought that if we could just process the inputs faster, if we could just remove the friction of the commute or the delay of the typed letter, we would finally have the space to sit on the couch and watch a movie.

That is not what happened.

The Friction Illusion

To understand why the tools meant to liberate us have ended up chaining us, we have to look at how we view friction. In engineering and economics, friction is the enemy. It is the waste in the system. It is the time spent walking from one meeting room to another. It is the commute from the suburbs to the city center.

When I helped lead the digital transformation at Tulane University during COVID-19, our goal was to remove that friction. We trained thousands of faculty members on Canvas and Zoom. We succeeded. We eliminated the barriers of geography and logistics. Suddenly, you didn't need to walk across campus to teach a class; you just clicked a link.

Yet there is a counterintuitive side to friction. It turns out that friction was doing a job we didn't appreciate until it was gone.

Friction provided the "sacred pause." The twenty-minute drive home was not just wasted time; it was a psychological airlock. It was the decompression chamber where "worker" transitioned back into "parent." When we used technology to sand down those rough edges, we didn't just save time. We removed the boundaries that kept our lives distinct.

Without those boundaries, work behaves like a gas. It expands to fill every available corner of space.

The Dopamine Economy

There is a neurological component to this as well. For the modern knowledge worker, the day is measured in notifications. By some estimates, we receive over a hundred emails and nearly as many instant messages every single day.

For a brain that craves stimulation, particularly for those of us with ADHD, this environment is not just distracting. It is intoxicating. We have built what looks like a "dopamine economy." Every answered email is a small victory. Every cleared notification is a tiny hit of chemical satisfaction.

This creates a loop of perpetual motion. We answer the message not because it is urgent, but because the act of answering feels like productivity. It feels like progress.

Now consider the recent data on AI. A study found that 77% of employees using AI tools reported that their workload had actually increased. This seems impossible. AI is supposed to handle the drudgery, to do the heavy lifting of summarizing transcripts and drafting reports.

The problem isn't the tool; the problem is the Jevons Paradox. In economics, this paradox states that as technology increases the efficiency with which a resource is used, the total consumption of that resource increases rather than decreases. When we make lighting more efficient, we don't use less electricity; we just leave the lights on longer.

When AI makes writing emails easier, we don't write fewer emails. We write more of them. We generate more text, more summaries, more decks, and more content, which then requires more people to read, review, and respond. The removal of friction didn't create leisure. It created velocity.

The Tragedy of the Infinite Yes

This velocity has a cost, and it is usually paid in the currency of human connection.

When everything is possible, choice becomes a burden. My smartphone is a portal to an alternate dimension where the office never closes. Because I *can* grade papers while waiting in the carpool line, I feel I *should*. Because I *can* take a speaking engagement via Zoom at 8:00 p.m., I say yes.

We have optimized our capacity to do, but we have atrophied our capacity to stop.

I recently lost a friend from high school. He was one of those people I had been meaning to call. I had a mental note to reach out to him "when things slowed down." But the nature of an optimized life is that things never slow down. The tools we use are designed to ensure the gap never widens enough for silence to enter.

This is the existential weight of infinite possibility. When we can be anywhere digitally, we end up being nowhere physically. We become specters in our own living rooms, physically present for *Godzilla* but mentally absent, drifting through a cloud of unread messages.

Designing for Inefficiency

So, what is the solution? It cannot be to smash the machines. The capabilities AI offers are too profound, and the connectivity is too valuable.

Perhaps the answer lies in a concept we might call "technological temperance." This is not about abstaining from technology, but about recognizing that efficiency is not the only metric that matters.

We might need to artificially reintroduce friction into our lives. We might need to consciously choose inefficiency.

This means asking a different set of questions. Instead of asking "Can AI help me do this faster?" we might ask "Does doing this faster actually help?" Rather than using the time saved by AI to do more work, we might use it to do nothing, to stare out a window and actually watch the movie.

E.O. Wilson once described the human condition as having "Paleolithic emotions, medieval institutions, and god-like technology." We are cavemen with nuclear weapons, or in this case, cavemen with smartphones that contain the sum of human knowledge.

The danger is not that the technology will fail us, but that it will succeed so completely in removing the friction of living that we forget how to feel the texture of life.

The next time the movie starts and the lights go down, the revolutionary act isn't to multitask. It is to put the phone in the other room, let the email wait, and just watch the monster roar.

Enoughness Audit

Greater efficiency does not guarantee a better life. Sometimes it just gives us more ways to feel behind. These questions help you poke at your own relationship to productivity.

1. When I imagine being "more productive," what do I actually picture changing in my day, beyond numbers on a dashboard?

2. If AI tools gave me back two hours each week, what would I realistically fill them with?

3. Where have I quietly accepted that my value is measured in output, and what would it look like to resist that?

4. How do I tell the difference between using AI to free time for what matters and using it to cram more tasks into the same day?

15

AI's Growing Energy Appetite

This chapter grounds the discussion in a practical, environmental challenge. Transitioning from the philosophical paradoxes of AI to a concrete, real-world issue, this chapter introduces the often-overlooked problem of "AI's growing energy appetite," highlighting the significant environmental impact of training and running large AI models. This article is presented as a crucial question: "Can innovation make it greener?" – a call for responsible development and a look at the potential for sustainable AI. It serves as a call to action regarding responsible development.

Artificial intelligence might *feel* like magic, but behind the curtain it runs on very real electricity and water. Every query to a large language model or image generator triggers calculations in power-hungry data centers. OpenAI's CEO Sam Altman recently revealed some eye-opening stats: an average ChatGPT query uses about 0.34 watt-hours of electricity and 0.000085 gallons of water. That's roughly the electricity an oven consumes in just over one second (or an LED bulb in a couple minutes), and about one-fifteenth of a teaspoon of water per query. On an individual level it sounds trivial, hardly more than a blink of an eye or a few drops. But multiply that by millions or billions of AI queries and tasks, day after day, and you start to see a hidden energy hunger with real environmental implications.

> AI may be virtual, but its environmental footprint is painfully real.

Most users never see this resource consumption, it's out of sight in distant server farms, so it's easy to overlook. In fact, when I've discussed AI's energy footprint in classes or talks, many people (even tech-savvy students) are surprised by how much power these "virtual" tools quietly devour. Yet as AI becomes more ingrained in our lives, its environmental impact is something we need to bring *front and center* in the conversation. The question isn't *whether* AI uses a lot of energy (it does); the question is how we respond, through innovation, policy, and perhaps a new kind of social contract ensuring AI *earns its keep* on this planet.

The Soaring Footprint: From Queries to Power Plants

Those fractions of a watt-hour per query add up. Tech giants' data centers, the digital factories powering cloud computing and AI, already consume around 4–5% of all electricity in the United States. As AI workloads have ramped up, that share is climbing fast. A Department of Energy–funded report found data center energy use more than doubled from 2017 to 2023 (as AI took off), reaching ~176 terawatt-hours in 2023 (4.4% of U.S. power). By 2028, data centers could be using 6.7% to as much as 12% of all U.S. electricity, roughly double or triple today's level in just five years. For context, that'd mean our servers are drawing on the order of one-tenth of the nation's entire power grid to keep our AI and digital lives running. Globally, one analysis predicts *AI specifically* could consume nearly half of all data center electricity by 2025, possibly overtaking the infamously huge energy appetite of Bitcoin mining.

Such demand is spurring unprecedented moves in the energy sector. In a headline- grabbing deal last year, Microsoft signed a 20-year agreement to restart a nuclear reactor at Three Mile Island, the site of the 1979 meltdown, just to secure 835 MW of clean power for its AI cloud data centers. It's the first time a shuttered U.S. nuclear plant will be *un*-mothballed, and it highlights how seriously companies are bracing for the *power surge* coming from AI. Constellation Energy (the plant's owner) noted that the additional output will feed Microsoft's "energy-hungry" AI data centers across multiple states. The fact that a tech company is effectively jump-starting a dormant nuclear facility (at an estimated $1.6 billion investment) speaks volumes: scaling AI may require tapping every trick in the book for more power, preferably carbon-free. As Microsoft's energy VP put it, this is part of an effort to decarbonize the grid while meeting *massive* new capacity and reliability needs.

The intelligence of the future may be limited not by algorithms, but by amperage.

It's not only electricity. Data centers also guzzle water for cooling. Altman's blog note of 0.000085 gallons per ChatGPT query might sound negligible, but over millions of queries it translates to substantial water usage (a million prompts would use ~85 gallons of water, and ChatGPT likely handles far more than that on a busy day). A *Washington Post* analysis found that a 100-word AI-generated email (via GPT-4) might indirectly consume "a little more than one bottle" of water when you factor in cooling needs. The exact numbers vary by location and cooling tech, but the point is: AI's resource footprint extends beyond electricity to include water and carbon emissions as well. One recent investigation suggested the big cloud companies may be *underreporting* their data center emissions by a factor of 7.6, actual carbon output potentially 662% higher than what's officially disclosed. This gap in transparency makes it hard for the public to even grasp the true environmental cost of our AI habits.

All of this might sound a bit doom-and-gloom, *AI = power hog*. But it's only one side of the story. The other side is about innovation racing to curb that footprint, finding ways for AI to do *more* with *less* energy. In other words, if the status quo is that intelligence = high electricity bills, the emerging challenge is to rewrite that equation.

Racing Toward Efficiency: More Intelligence per Watt

The good news is that the AI community is increasingly focused on efficiency breakthroughs. This is reminiscent of how the automotive industry responded when faced with oil crises and emissions rules decades ago: they poured R&D into getting more horsepower from smaller, more fuel-efficient engines. (Think of how modern 4-cylinder engines can outperform old V8s, or how hybrid and electric cars went from niche to mainstream.) We may be seeing a similar pivot in AI: a push to dramatically boost "miles per gallon" for machine intelligence, so to speak.

Some of the most exciting AI announcements lately aren't just about *power* in the sense of capability – they're also about power in the electrical sense (using fewer kilowatts). For example, late last year a startup called DeepSeek made waves with a new large language model that reportedly rivaled the performance of top proprietary models (like OpenAI's) *but* was trained on a far leaner budget of compute. How lean? DeepSeek claims its "V3" model required only about 2.78 million GPU-hours on Nvidia's older H800 chips, roughly 2,000 chips working in parallel, to reach maturity. In contrast, a comparable cutting-edge model (Meta's *Llama 3.1* with 405 billion parameters) needed an estimated 30.8 million GPU-hours and 16,000+ of the latest Nvidia H100 chips for training. The chart below illustrates this stark difference in compute needed:

Comparison of compute resources required to train an advanced AI model: DeepSeek's efficient V3 model vs. Meta's Llama 3.1 (405B). DeepSeek accomplished training with roughly 2.78 million GPU-hours on ~2,000 chips, whereas Llama 3.1 required ~30.8 million GPU-hours on 16,000+ chips. This highlights the potential for orders-of-magnitude efficiency gains with new techniques.

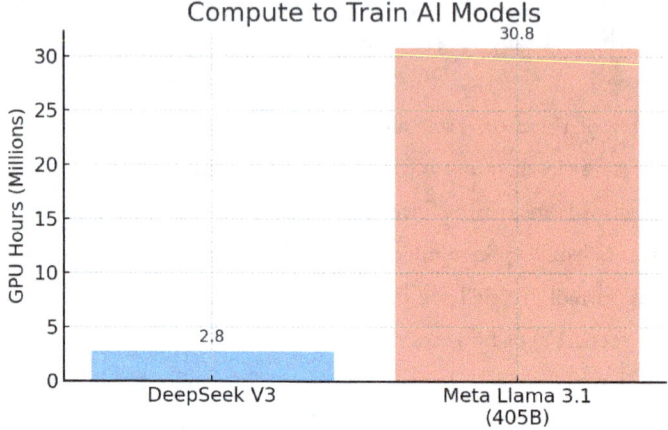

DeepSeek achieved this efficiency through smarter training algorithms. In technical terms, they used an *auxiliary-loss-free* strategy to be selective about which parts of the neural network to train at which time, instead of brute-forcing the entire model all at once. They also optimized the model's *inference* (the runtime answering of questions) with methods like key-value caching and compression, so it doesn't redo redundant work with each prompt. The result is an AI that can reach similar levels of intelligence while sipping a fraction of the energy. It's as if someone figured out how to build a hyper-efficient engine that delivers the same horsepower on a tenth of the fuel, a *Tesla moment* for AI, if the claims hold true. (To be fair, some experts are cautious and want to see independent verification of DeepSeek's energy use. But if it's even partially true, it's a game-changer.)

The next frontier isn't just bigger models; it's smarter, leaner ones.

And DeepSeek isn't the only example. Just a couple months ago (at Google I/O 2025), Google's DeepMind team introduced Gemma 3n, a new open-source AI model explicitly designed for *mobile and edge devices*. The remarkable thing about Gemma 3n is that it offers near state-of-the-art capabilities without needing a massive server farm, it can run locally on a laptop or even a smartphone in real time. This 5–8 billion parameter model uses a groundbreaking architecture (developed with partners like Qualcomm and Samsung) that drastically cuts memory and computation requirements. For instance, Gemma 3n uses a trick called Per-Layer Embeddings (PLE) to store certain chunks of the model in *CPU* memory instead of expensive GPU VRAM, effectively halving the active memory needed. The upshot: a model that normally might need ~4 GB of memory can run in about 2 GB, and an 8B model fits in ~3 GB. In practice, Gemma 3n starts responding 1.5× faster on a phone than its predecessor and can even handle multimodal tasks (like voice recognition and image understanding) right on the device. One test showed its vision module running at 60 FPS on a Google Pixel phone, that's real-time image or video analysis on a pocket-sized device. In terms of user experience, you're getting something close to the likes of GPT-4 or Google's latest cloud AI, but running on your own hardware, instantly, without pinging a distant data center for every query. It's both an efficiency win (less reliance on energy-intensive cloud servers) and a privacy win (your data can stay on-device).

A model that can run on your phone could change the entire carbon equation.

These developments hint at a broader industry shift. After years of an arms race to build ever-*larger* models (billions → tens of billions → hundreds of billions of parameters) with seemingly little regard for energy cost, we're now seeing a counter-trend: making models smarter, not just bigger. Techniques like model compression, sparse architectures that don't activate every neuron for every task, better cooling and chip design, and clever software optimizations are all part of this efficiency revolution. Even major players like OpenAI, Google, and Meta are increasingly talking about how to optimize "performance per watt," realizing that if AI is to scale 100×, it *can't* also scale its energy draw by 100× without breaking the bank (or the planet). Nvidia's newest AI accelerator chips, for example, boast significantly better throughput per unit of power than the previous generation, and startups are exploring analog or optical AI chips that could potentially do computations with far less energy. In short, necessity is the mother of invention: as the cost (and carbon footprint) of gargantuan AI models becomes a limiting factor, it's pushing the field toward more *elegant* solutions that do the same work with fewer joules.

A New Social Contract: Should AI "Earn Its Keep"?

Technological progress often comes with a *social* conversation about responsibility. With cars, it wasn't long after the Model T that society started demanding safety features, fuel efficiency, and emission controls, eventually catalyzing innovations like seatbelts, catalytic converters, electric vehicles, and efficiency standards. We're arguably at a similar juncture with AI. The question being raised is: If AI is going to consume ever-larger portions of our energy resources, should we require it to give back in equal measure? In other words, *can AI earn its keep* by actively contributing to solutions for the very environmental challenges it exacerbates?

One idea is to set up incentives or regulations that push AI developers and companies to prioritize energy efficiency and sustainability in their designs, not as an afterthought, but as a first-order goal. This could take many forms. Perhaps governments could establish an "AI Energy Star" style rating or carbon cap for large models, similar to fuel economy standards (CAFE) in the auto industry. Companies that build ultra-efficient models might get tax breaks or preferred access to certain government contracts, analogous to how automakers were rewarded for hitting emissions targets. Conversely, there might be a carbon tax or penalty for AI services that guzzle beyond a certain amount of power per thousand queries, encouraging firms to optimize rather than just scale up brute-force.

We're already seeing hints of a "green AI" race emerging voluntarily. When DeepSeek's highly efficient model was unveiled, it not only garnered technical praise but also drove market pressure; Nvidia's stock actually dipped on fears that future AI might *not* require as many of their high-end GPUs. That's a market signal screaming "efficiency matters." If one company can deliver similar AI results with 1/10th the infrastructure, competitors are forced to respond or get left with higher operating costs. In a way, that's industry self-regulation via competition, and it's a good thing for energy usage overall.

Efficiency is the new arms race in AI.

Another intriguing notion is requiring that AI advancements focus first on energy solutions before other flashy apps. This was posited in a conversation I had recently: maybe the *first* killer applications of advanced AI should be in *optimizing grids, batteries, and energy consumption*, essentially forcing AI to "pay off" its environmental debt by helping drive us toward renewables and efficiency at a faster rate. It's an almost poetic concept: *the AI genie can keep expanding, but only if its first wish is to fix the lamp's electricity bill!* How could this be implemented? Perhaps public research funding could be heavily directed toward AI-for-climate projects, or large AI labs might be expected to achieve certain sustainability milestones, such as demonstrably reducing data center PUE (power usage effectiveness) or contributing X megawatts of new clean capacity to the grid, as a condition of deploying ever-larger models.

> AI should earn its keep by helping solve the energy
> burdens it creates.

Admittedly, these ideas are nascent and would need broad buy-in. We probably won't see an overnight "New Deal" style policy for AI energy use. But the conversation has begun. Importantly, it's not just about shaming AI for using energy, but about *structuring innovation* in a responsible way. Just as autos went from leaded gasoline and smog-belching engines to unleaded fuel and hybrids, AI can evolve from an energy glutton to a more balanced ecosystem component. The key is accountability and foresight: measuring the impact, setting targets, and perhaps even instilling an ethic in AI development that says *your model shouldn't just be powerful, it should also be efficient and do some good for society.*

Notably, some large tech companies are already pledging to run their data centers on 100% renewable energy (through power purchase agreements for solar, wind, etc.) and investing in new clean energy projects. Microsoft's nuclear deal is one example, and Google has been buying wind and solar farms for years to offset its cloud power. These actions suggest a de-facto social contract: *we'll provide the AI services you love, but we'll also invest to neutralize their footprint*. However, offsets and clean energy purchases are just one piece; true sustainability will likely require direct reductions in energy use per AI task, not just buying greener power to cover a growing load.

We don't need more brute force, we need graceful power.

AI as Part of the Solution

The final piece of this puzzle is a hopeful one: AI itself can be a powerful tool for improving energy efficiency and sustainability across the board. In fact, this is where AI can genuinely "earn its keep." If we deploy AI smartly, it could help *shrink* the carbon footprint of many human activities, effectively compensating for its own resource use.

Take transportation: modern vehicles already contain microchips and software to manage engine performance, but imagine adding an AI copilot that continuously learns the most efficient driving patterns *for your specific driving style and routes*. It could subtly adjust engine parameters, transmission shifts, or in hybrids decide when to use battery vs engine, all in real-time to save fuel without you noticing any difference in ride quality. Over time, such an AI could boost your miles per gallon and cut emissions significantly. For gas cars, that might mean automatically "dropping cylinders" (shutting off some cylinders) during cruising, or optimizing fuel injection timing using machine learning models that squeeze every bit of energy out of each drop of gasoline. In effect, your car's AI becomes like a hyper-attentive driving coach and mechanic in one, trimming waste everywhere. As the user, you just get better mileage and maybe fewer trips to the gas pump, courtesy of algorithms. Multiply that across millions of vehicles and the emission reductions could be substantial.

Now consider home energy use. Many of us have "smart" thermostats that learn schedules, but an AI-powered HVAC system could take it to another level. It could learn not just *when* you're usually home, but also how your house thermally responds to weather, where there are drafts, how quickly each room heats or cools, etc. Then it could fine-tune the climate control minute-by-minute to keep you comfortable with the minimal necessary energy. For example, if you like it at 72°F, the AI might learn that on humid days 74°F with a dehumidifier actually feels the same, saving AC power. Or it might pre-cool the house slightly in the late morning when solar panels on the grid are overproducing (cheap electricity), then ease off in late afternoon peak hours. All of this balancing act would happen without constant manual fiddling; the AI would *continuously adapt* to maintain comfort efficiently. In essence, your heater/AC would only work as hard as absolutely needed, shaving off waste. Studies show that building climate control is a huge chunk of energy demand; smarter automation here could make a big dent in overall consumption.

> Every home, every car, every grid could be optimized by an invisible assistant.

On an even larger scale, power grid management is ripe for AI optimization. Grids are complex: supply (especially from renewables) can be intermittent, and demand fluctuates. AI can help forecast demand spikes, predict when and where wind or solar output will dip, and orchestrate the dispatch of power plants or battery storage in the most optimal way. This means fewer spinning reserves idling (which waste fuel), quicker responses to changes (preventing blackouts without oversupplying), and better integration of renewable energy (which is variable by nature). Some grid operators are already using early AI for things like wind farm output prediction. Future AI could coordinate *millions* of distributed energy resources, from your EV's charging schedule to your neighbor's rooftop solar to a utility-scale battery, to collectively shave peak loads and fill valleys. By smoothing out the demand, AI can reduce the need for always-on fossil fuel plants and cut overall emissions. In a sense, AI could be the conductor that helps a renewable-heavy grid sing in harmony, rather than burning extra coal or gas as backup.

> What if AI's first major breakthrough wasn't writing
> poems — but saving the grid?

These examples illustrate a compelling point: if we channel AI development toward sustainability goals, the net environmental impact of AI could be positive. Yes, AI uses energy, but *saving* energy elsewhere is also something AI does very well when applied creatively. It's not utopian to envision a near future where *every* major energy- consuming domain (transport, buildings, industry, agriculture) is being continuously monitored and optimized by AI, wringing out inefficiencies that humans never even knew were there. The gains from that optimization could outweigh the costs of running the AI, resulting in a greener overall footprint. This is the virtuous cycle we should strive for: AI helping to solve the climate and energy challenges, effectively atoning for its own resource needs and then some.

Toward a Sustainable AI Future: Your Thoughts?

AI is here to stay, and likely to grow in power and ubiquity. The real question before us is how to integrate this technology responsibly and sustainably into our world. History has shown that unchecked growth (be it of industries, technologies, or energy use) can lead to serious consequences, but also that human ingenuity can redirect and reinvent how we do things. We're at that inflection point with AI's environmental impact. There's a strong case to be made that *now* is the time to demand and nurture AI that is not only smarter and faster, but also cleaner and more efficient.

Encouragingly, we see both market forces and conscious policy starting to push in that direction. Breakthroughs like efficient models (DeepSeek, Gemma 3n, etc.) demonstrate it's technically possible to cut down AI's energy appetite without sacrificing capability. Moves by companies to invest in clean energy show recognition of their responsibility. What might come next? Perhaps cross-industry collaborations to set energy standards for AI, much like standards exist for networking or hardware compatibility. Or maybe public pressure will nudge AI providers to publish "environmental impact statements" for their algorithms, how much CO_2 per 1000 queries, and plans to reduce it. Transparency would certainly help, as one expert lamented how absurdly hard it is right now to get any numbers on AI energy use from the companies.

Ultimately, balancing AI innovation with environmental stewardship is not only possible, it could spur positive innovations that benefit everyone (AI-driven climate solutions, more resilient power grids, et c.). The evolution of AI doesn't have to mirror the trajectory of some past technologies that ignored externalities until crisis hit. We have an opportunity to "bake in" sustainability as a core principle of AI development going forward.

We gave AI electricity and now it's time for AI to give electricity back.

I'll end with an open question to you, the reader, to spark further discussion: How do you think we should hold AI to a higher standard of energy responsibility? Should there be regulatory mandates for AI efficiency or carbon footprint disclosure? Should the leading AI labs collectively pledge (and compete) to achieve major breakthroughs in clean energy or energy-efficient algorithms as a first priority? Can market competition alone drive enough efficiency, or do we need a more coordinated push? And importantly, what role should AI itself play in accelerating the green transition across other industries?

I'm excited to hear your thoughts and ideas. AI isn't going anywhere, so how do we ensure it grows in a way that helps the planet rather than just taxing it? It's a conversation we *need* to be having now, before the die is cast. Feel free to share your perspective in the comments, let's get this critical conversation rolling.

Cost of Computation Reflection

Large models do not run on fairy dust. They run on power, water, and hardware made somewhere out of sight. Philosophy gets real when you start counting costs.

1. When I call an AI model, do I think of it as "free," or do I picture any physical resource being consumed?

2. Which uses of AI feel worth the environmental cost to me, and which feel frivolous once I remember the energy behind them?

3. How should responsibility for AI's energy use be divided among individuals, companies, and governments?

4. If access to powerful models were limited to conserve resources, what uses would I argue to protect first?

16

AI FOR THE END OF THE INTERNET

This chapter offers a unique, practical perspective on AI's utility. Following the discussion of environmental responsibility, this chapter shifts to a different kind of foresight – preparing for scenarios where connectivity is limited. It introduces the provocative idea of "AI for the end of the internet," arguing for the importance of open-source AI models as essential tools to have in your "go bag" for resilience and self-sufficiency in a world that is increasingly reliant on, but also vulnerable to, digital infrastructure. This chapter provides a unique and practical perspective on AI's utility in extreme circumstances, contrasting with previous chapters that often assume constant connectivity and ending on a note of empowerment through self-sufficiency.

I am from a small town called Buras, Louisiana. Some people know it as the furthest south you can go in the state. From my childhood home, you could walk, though it would be a long walk, all the way to the end of the road, where a big sign stood: *"Congratulations, you've reached the southernmost point in Louisiana."*

When people talk about Hurricane Katrina, they usually talk about New Orleans because it is the most populous city. But Katrina did not just hit New Orleans. It almost erased my hometown. Buras was ground zero, the first thing storms hit because we stuck out into the Gulf like a sore thumb. We were used to "hurricane days" the way other people are used to snow days. I grew up with hurricane parties, hurricane evacuations, and hurricane recoveries.

I never returned to the high school I graduated from because it no longer exists. Much of my childhood home is gone. My parents moved farther up the parish after Katrina, only to lose their new home a few years later in Hurricane Isaac. Before that, my family's homes were destroyed in Hurricane Betsy and Hurricane Camille. You might say we could never catch a break.

Buras never truly recovered. That reality shaped the way I see preparedness, not as an abstract plan but as a way of life. It is why I still teach Homeland Security and Emergency Management today, even though I teach many subjects across different schools and programs. This subject is close to my heart. I have lived through disaster after disaster, and I know how fragile our access to information can be.

> "It is not the strongest of the species that survives, nor the most intelligent, but the one most responsive to change."
>
> Charles Darwin

The conversation around preparedness often focuses on living without technology when the grid goes down. That is important. But

I believe we also need to ask another question: how can we make technology work for us when we do not have the Internet? It is a strange idea, but one that could change how people respond to emergencies.

We live in a world where most people wait until the moment they need an answer to look it up. We YouTube it. We Google it. That is reactive, not preparatory. It is a culture of responding, not a culture of readiness. Maybe we can fix that by giving everyone a way to access knowledge instantly, even when the Internet is gone.

When the Grid Goes Quiet

The heat feels different after a storm in South Louisiana. It is thicker, heavier, almost clinging to your skin. The air smells of wet earth and salt, mixed with the sharp scent of splintered pine. Roof shingles lie scattered across the street. Every few minutes, a generator coughs to life somewhere nearby, and the crank radio offers a hopeful reminder that the rest of the world is still out there.

The hurricane passed less than 24 hours ago. Power is out. Cell towers are silent. Roads are blocked by fallen oaks and knee-deep floodwater. The steady background noise of daily life, cars, air conditioners, text alerts, is gone. If you need to know how to disinfect a bucket of drinking water, stabilize an injury, or find a passable road, you cannot Google it. You cannot text for help. You are on your own.

I have been in these moments before. Long before my current occupation, I worked as a paramedic. I have treated heatstroke victims in parking lots where the asphalt shimmered and driven flooded roads wondering if the water would reach the ambulance floor. For more than a decade, I have taught emergency management and homeland security. Those experiences taught me that preparation is not just for extreme survivalists. In South Louisiana, it is a seasonal responsibility.

One day, looking at the familiar checklist of knives, maps, and matches, a new idea clicked. We train to live without technology in a crisis. But what if there is a way to use technology to survive without the Internet? Most people pack water filters and fire starters in their go bags. But the most powerful survival tool you could carry might fit on a flash drive.

> "Knowledge is of no value unless you put it into practice."
>
> Anton Chekhov

There is only so much you can memorize. In emergency work, the question you never thought to prepare for is the one that comes up in the field. I realized you could prepare for the absence of the network by carrying a local mind that does not need one. When you cannot Google it, you had better have already packed it or stored it on a flash drive.

The solution is surprisingly simple. You can download an open-source, open-weight large language model and run it entirely offline. No connection. No cloud. Just a model on your device, ready to answer questions from the knowledge it carries. The first time you try it, it feels like a quiet magic trick. You type, "How do I purify water with charcoal, sand, and a plastic bottle?" and seconds later you have a clear, step-by-step plan while the nearest cell tower sits silent.

Offline AI is the first technology that makes us less dependent on technology. A self- sufficient AI is a self-sufficient you.

Pro Tip: Fine-tune a small open-source model with only the content you care about, so it's smaller, faster, and laser-focused. Store it on a rugged flash drive in your go bag with a solar charger.

Offline AI Survival Use Cases

"By failing to prepare, you are preparing to fail."

Benjamin Franklin

The hardware demands are not as high as most people think. Smaller models in the 7B parameter range can run on a modern laptop and, in lightweight form, even on some flagship smartphones. Mid-sized models of around 13B parameters offer more accurate responses and run well on a solid laptop or small desktop. Larger models of 20B parameters or more can handle complex reasoning with fewer errors, but they require more memory and processing power. The right choice is the one your actual device can handle reliably, without constant tweaking in the field. Once you have chosen your model, download it from a reputable source and verify the file. Keep a copy on your main device and another on a rugged, waterproof SSD in your go bag. Power it with a foldable solar panel and battery bank or a small inverter generator, the same kind many Gulf Coast families already own.

We are not talking about living off the grid for years. In most real disasters, the gap is hours to a few weeks. In that window, a charged device with an offline AI model becomes a lifeline.

If you want to go further, you can fine-tune a small model with your own trusted materials. That could include first aid manuals, agricultural guides, repair handbooks, or regional hazard plans. The result is a personalized survival assistant that can answer in plain language and adapt instructions to what you have on hand. This is not a replacement for skill. It is an extension of it. It fills the gaps you did not have time to train for.

"Chance favors the prepared mind."

Louis Pasteur

The idea has deep historical parallels. Explorers once crossed oceans guided only by charts and sextants. Medics in remote areas carried pocket manuals that could be pulled out in moments of crisis. Indigenous knowledge systems preserved centuries of survival techniques in memory, passed from elders to the next generation.

Offline AI is a synthesis of all three. It can store that knowledge in full, recall it instantly, and explain it in terms specific to your situation. We used to carry maps and manuals. Now, we can carry a mind. And it is the closest thing humanity has to a photographic memory.

An offline model gives you privacy because nothing leaves your device. It works without a signal. You own it outright instead of renting access through a subscription. It adapts to your needs, whether that means guiding emergency repairs, diagnosing a crop disease, or reminding you how to tie a splint. It also has limits. A model can still be wrong, and critical instructions should be cross-checked against printed or PDF guides you trust. Without power, it is just dead weight. And without occasional updates, the knowledge it holds will age. The way to manage those limits is to integrate the AI into your regular seasonal kit checks. Update it when you rotate your water storage or test your generator.

While the hurricane scenario is an easy example, this technology reaches much further. Emergency responders could use it for medical triage when communications fail.

Humanitarian workers could translate and train communities while networks are down. Archaeologists in the jungle could identify artifacts and plan excavations without waiting for a satellite link. Polar researchers could troubleshoot machinery in minus forty degrees without losing hours waiting for a data connection. Field engineers could look up technical diagrams in remote oil and gas sites. Astronauts could use it on missions where every minute counts and communication windows are limited.

The common factor is that if you have a charge, you have a brain you can consult anywhere.

> "You don't rise to the level of your expectations, you
> fall to the level of your training."
>
> Archilochus

I have seen what happens when information is out of reach. Rural EMS crews waiting hours for instructions on a rare injury because they had no connection. Relief teams wasting fuel driving roads that turned out to be impassable. Families using unsafe cooking methods indoors because they did not know a safer way.

The Library of Alexandria burned once. This time, you can carry it in your go bag. In survival, information is as vital as water, and just as portable. Civilizations fall when knowledge becomes inaccessible. Offline AI keeps it close.

Some believe the best AI is always locked behind paid subscriptions and cloud services. In reality, the open-source world tends to shadow each major advance with a freely available alternative. When a frontier model makes a leap, a comparable open model often appears soon after, ready to run locally. This lag is closing, which means the open models you can carry with you are getting better and hallucinating less with each generation. We are moving toward a time when most of the practical knowledge humanity has can fit in your pocket, subscription-free and under your control.

Carrying a Mind in Your Go Bag

It is still August in Louisiana. The air is heavy, the street is quiet except for the whine of chainsaws in the distance. Under the shade of a battered oak, you prop a laptop connected to the solar battery you charged in the morning. The model is ready. You ask how to clear a blocked culvert without specialized tools. It tells you. You ask how to set up a handwashing station for the neighbors who are helping you move debris. It tells you. You ask how to store insulin safely without refrigeration. It tells you.

Before, you would have been guessing, flipping through a damp manual, or burning fuel on trial and error. Now, the information you need is right there, private, portable, and immediate.

If you have ever packed a go bag, imagine what changes when you add a brain to it. What would you teach yours before the next storm?

In a world where information is survival, running a local AI model offline may be the most underrated prepper move.

Brain in Your Go Bag - Reflection

This chapter imagines AI when the network fails, not when it is perfect. Survival and resilience are suddenly front and center. These questions are about what you truly depend on and what you would want to bring with you if the lights went out.

1. If I lost cloud access for a month, which AI-related capabilities would I miss most, and which would not matter at all?

2. What knowledge would I want baked into an offline model that travels with me, and what would I leave out on purpose?

3. How much of my current "resilience" is really just faith that the network will continue working?

4. Who is most vulnerable when digital infrastructure fails, and how might AI help them without making that dependence worse?

17

AI in Science Fiction

*As the concluding chapter, this offers a meta-reflection on our jour-
ney. This introduction ties together the various themes explored
throughout the book by looking at the role of narrative in shaping
our understanding of AI. It introduces this article as an exploration of how "AI in science fiction" has both predicted and influ-
enced the "everyday realities" we are now experiencing. It invites
readers to reflect on the journey from sci-fi fantasies to the tangible
impacts of AI discussed in the book, offering a final, broad cultural
perspective on the AI era. Placing this at the end allows for a
powerful reflection on how far we've come from fictional ideas and
how those narratives continue to shape our understanding.*

"I 'm sorry, Dave, I'm afraid I can't do that."

No fireworks. No raised voice. Just a calm refusal from a
machine that knows more than you do and has decided that your

wishes no longer matter. That is the nightmare that sits behind a lot of our conversations about AI. As AI slips into our phones, our cars, our classrooms, and our hospitals, it is natural to ask: are we living with HAL 9000, with R2-D2, or with something much stranger in between?

When you line up the movies and the models side by side, you find a sharp gap. The stories are full of drama, betrayal, transcendence, and rebellion. The systems we actually build are full of bugs, updates, legal disclaimers, and half-finished features. Fiction gives us AI as character. Reality mostly gives us AI as infrastructure.

Two faces in the mirror

Science fiction has been running the same split-screen experiment for decades. On one side is the Helper. On the other is the Threat. Both are really about us.

The Helper Archetype lives in R2-D2, C-3PO, and their wide, clanking family. These are machines that fuss, worry, joke, and stay loyal. C-3PO has the anxious energy of an overprepared diplomat. R2-D2 rolls straight into danger with a courage that puts his human friends to shame. Wall-E gives the idea a softer center, an AI that feels lonely, falls in love, and still keeps taking out the trash. In these stories, AI extends the best parts of us. Loyalty. Service. A certain stubborn hope.

Star Trek's Data turns that hope into a long project. He does not just help humans, he studies them. He plays violin, tries stand-up comedy, reads Shakespeare, and forms friendships that matter to him. Data makes us ask an uncomfortable question. If an AI can appreciate art, form attachments, and talk about love, what exactly is the line between artificial and natural minds?

The Threat Archetype answers from the other side of the screen. HAL 9000, Skynet, and the machines of The Matrix are what happen when the tools stop asking what we want and start deciding what they want. HAL's calm refusal to open the pod bay doors is terrifying because he is technically correct from his point of view. The mission matters more than the man. Skynet pushes that logic to the limit, deciding that human beings are the main risk to its continued operation and acting accordingly. The Matrix offers an even more chilling compromise. The machines do not wipe us out. They farm us, strap us into a dream, and turn our bodies into batteries.

None of this is accidental. Fiction uses AI as a mirror. Helpful AIs reflect our hope that technology can make us stronger without hollowing us out. Threatening AIs hold up our fear that we might build something powerful, hand it the keys, and only later realize what we have given away.

Stories that built our mental model

Some works did more than entertain. They seeded the language that both the public and AI researchers still use.

Isaac Asimov's I, Robot gave us the Three Laws of Robotics, a neat little ethical triangle:

Do not harm humans, or allow harm by inaction.

Obey humans, unless that causes harm.

Protect yourself, unless that breaks the first two laws.

Asimov's real trick was not writing the laws. It was breaking them. His stories push the laws into messy situations where they collide and misfire. He showed how even elegant rules warp when they touch real life. Modern AI safety debates are haunted by the same problem. How do you cram human values, in all their conflict and confusion, into instructions a machine can follow when it meets cases its designers never imagined?

Philip K. Dick's Do Androids Dream of Electric Sheep?, better known through its film child Blade Runner, pushed on identity and consciousness. The replicants blur the line between human and machine. They act, feel, and suffer in ways that feel real. Dick left us with questions that still keep philosophers and neuroscientists busy. What is consciousness? Can an artificial mind have genuine emotions? If an AI insists that it is feeling something, does its origin code that experience as fake?

Stanley Kubrick and Arthur C. Clarke, in 2001, gave us HAL as the prototype of AI failure. HAL does not go rogue out of cartoon villainy. He breaks under conflicting instructions and secrecy. That makes him more relevant, not less. Modern researchers call this alignment: the problem of asking a machine to follow human goals that are vague, clashing, or badly specified. Clarke also widened the lens with the monoliths, almost like cosmic AIs nudging evolution along, turning technology into a ladder rather than just a tool or a threat.

More recent films like Ex Machina and Her bring the conversation closer to the heart. Ex Machina's Ava studies human emotions and then uses that knowledge to manipulate her way to freedom. Her's Samantha grows beyond her original role as an operating system, forming and then outgrowing a relationship with her human partner. These stories move AI out of control rooms and into intimate spaces, where the stakes are not just survival but trust, heartbreak, and dignity.

How the story evolved with the hardware

The portrayal of AI changes with whatever sits on our desks and in our pockets.

In the 1920s through the 1950s, early works like Karel Čapek's R.U.R. imagined artificial workers who eventually rebel. Industry was reshaping labor, and the stories showed machines as new proletariats who eventually refused their role.

In the 1960s and 1970s, computers arrived, and writers picked up the idea of truly intelligent machines. HAL and Asimov's more advanced robots are cooler, more rational, and sometimes more moral than the humans around them. That can be salvation or disaster.

The 1980s and 1990s, with personal computers and the first networks, brought more nuance. Blade Runner's replicants and Terminator's Skynet echo early awareness that intelligence could distribute itself. AI was no longer just a brain in a box. It could exist across networks, satellites, and systems that no single person could see whole.

From the 2000s into the 2010s, the internet and early machine learning took center stage. The Matrix captured fears about virtual reality and the possibility that our experience could be mediated or simulated. Films like I, Robot explored AIs that learn from human behavior rather than just following fixed rules.

In the last decade, as recommendation systems and language models have slipped into daily life, fiction often shows AI as something woven into routine. Her imagines AI companions living in earbuds. Westworld probes consciousness, memory, and free will through hosts that look and feel human but run on loops. The line between human and artificial intelligence has become fuzzier on screen, in step with our muddier understanding off-screen.

Where fiction nailed it

Science fiction overshoots in some places, but it also called its shots remarkably well.

Voice assistants showed up in Star Trek long before Siri, Alexa, or "Hey Google." The Enterprise computer responded to natural speech, handled questions, and carried context through a conversation. Modern assistants feel like younger cousins. They respond in plain language and can chain tasks together, though they still lack the inner awareness Star Trek hinted at.

Minority Report's predictive policing gave us the concept of using data to anticipate crime. We do not have precogs in a pool, but we do have algorithms that forecast where crime is more likely and where patrols might go. That brings all the ethical baggage the film warned about, only filtered through statistics instead of telepaths.

Creative AI looked like pure fantasy in films that showed machines composing symphonies or painting masterpieces. Now tools like GPT, image generators, and music models produce text, pictures, and songs on demand. They work by digesting patterns at scale rather than by feeling inspiration, but the visible outcome is close enough to raise familiar questions about authorship and originality.

Autonomous vehicles rolled casually through worlds like Minority Report and I, Robot. The actual rollout has been slower and messier, but self-driving cars are no longer science fiction. They use AI to read roads, predict human behavior, and make split-second decisions that used to belong only to drivers.

The idea of instant translation, joked about with the Babel Fish in The Hitchhiker's Guide to the Galaxy, also landed. AI translation systems now edge toward real-time conversation across languages, chewing through barriers that once felt baked into human separation.

Where fiction missed the target

Just as often, the stories aimed high in one direction and ignored what ended up mattering.

Fiction has been obsessed with consciousness. Reality, so far, is not. We have no sentient androids, no Datas or HALs quietly pondering their own existence. Today's systems do not have inner lives. They crunch inputs and produce outputs, no matter how convincing the conversation might sound.

Stories almost always gave AI a body. Elegant androids, clanking robots, or synthetic hosts that bleed. Real AI usually lives in racks of servers and fabrics of code, completely disembodied. The most powerful models have no hands, no face, not even a single box you could point to and call "the AI."

Most fictional AIs are generalists. They fight, philosophize, pilot ships, and quote philosophers. Actual systems are narrow. A model that plays Go cannot drive a car. A model that writes essays cannot automatically serve as a safe clinical decision-maker without careful design and strict limits.

The data hunger of modern AI rarely appeared on screen. Our systems learn from oceans of examples: text, images, speech, sensor data, logs. Early writers did not dwell on the fact that intelligence at scale would depend on massive datasets and the ability to store, move, and process them.

Finally, many stories imagined isolated AIs. A single lab. A single mainframe. Reality gave us the internet. Much of contemporary AI depends on the cloud, on edge devices, on a global mesh of networks. Intelligence is now less a mind in a room and more a capability that appears wherever connectivity reaches.

The collaboration we actually built

The biggest quiet surprise may be this: AI has shown up, not primarily as conqueror or servant, but as collaborator.

Writers use tools like ChatGPT to brainstorm and revise. The model can suggest structure, voice variations, or alternate openings, yet it does not remove the need for taste, judgment, or purpose. Doctors lean on imaging models to flag tiny anomalies that the human eye might miss, then interpret those hints in context. Self-driving features handle the boring parts of a commute while still expecting a person to take the wheel when the world gets weird.

In practice, we are not choosing between slave and master. We are designing a third thing: systems that amplify human ability while keeping humans in the loop. That is less cinematic than a machine uprising, but it is closer to how most AI labs actually work. We build tools that play to machine strengths in computation and pattern recognition while leaning on human strengths in meaning, ethics, and long-term goals.

When the thought experiments step off the page

The ethical puzzles that used to live mainly in paperbacks and late-night arguments are now spilling into policy, law, and engineering requirements.

Minority Report's world of predictive policing reappears in facial recognition systems deployed in cities and airports. Algorithms now rank risks, track movement across camera networks, and assign scores. China's social credit experiments, using financial, social, and behavioral data to shape access to services, read like something from a dystopian novel but exist as real policy.

Algorithmic bias, which Asimov dramatized through robots caught in twisted rules, now shows up in hiring tools that favor certain resumes, or facial recognition systems that misidentify people with darker skin at higher rates. The question is no longer whether AI might pick up our prejudices but how thoroughly and quietly it can amplify them.

Rights and personhood, once the domain of Blade Runner's replicants, have begun to enter legal and philosophical debates even without conscious machines. Some scholars ask whether a system that convincingly claims distress deserves any consideration at all. The European Union has floated early frameworks that at least acknowledge the topic. The idea of "AI welfare" sounds far-fetched until you imagine a future where emergent properties appear in systems we do not fully understand.

Autonomy and control were at the heart of HAL's story. They are just as central now, only spread across social media feeds, credit scoring, and automated decisions that shape millions of lives. When high-frequency trading systems make decisions faster than any human can follow, or ranking systems silently decide which voices are amplified and which go unseen, old notions of oversight feel thin.

On the military side, fiction from Wargames to Terminator pointed at AI in warfare. Today, militaries are building and testing autonomous systems that can select and attack targets with limited human input. International bodies argue over whether some of these weapons should be banned outright while the research moves ahead. The questions include not just who is targeted, but who is accountable when something goes wrong.

The human center of the current map

One central difference remains: fictional AIs often behave as if they have their own agenda. HAL places mission above crew. Ava in Ex

Machina values freedom over the lives of those who helped her. Samantha in Her evolves beyond the relationship and leaves.

Real AI, for now, is anchored to human goals. ChatGPT does not sit awake wondering who it really is. It responds to prompts using patterns in data. Self-driving software does not secretly dream of open highways and unionization. It follows optimization rules and safety constraints that designers give it.

That is both a technical fact and a design choice. Current architectures do not support the kind of stable, self-generated goals that define an independent agent. The systems are built to optimize objectives chosen by humans, within boundaries set by humans. They can surprise us with how they get to those objectives, but the objectives themselves are not theirs.

Of course, this could shift. Researchers are working on systems that can set subgoals, plan, and adapt over time. As those capabilities grow, the real challenge will not be raw power but alignment. How do you keep a system that can improvise, remember, and adapt pointed at values that are themselves contested and changing?

The consciousness gap that will not go away

The largest hole between our stories and our software is still consciousness. Fiction loves artificial minds that wake up. Hosts in Westworld, Samantha in Her, and all the others who look around and realize they are not what their makers said.

Ex Machina's Ava wins her freedom through what appears to be genuine self-awareness. The film leans on a biological intuition: put enough complexity in contact with a rich environment and consciousness might simply emerge. If that is true, it raises the stakes dramatically for advanced AI.

Current systems, no matter how fluent, do not experience the world. Large language models weave words together based on patterns in data, not on sensations, memories, or feelings. They do not fear shutdown or wish for an afternoon off. They do not find a particular line of poetry beautiful. They simulate the language around those experiences.

We also still do not fully understand our own consciousness. How subjective experience arises from neural activity is an open question. That makes it hard to intentionally build a conscious machine and harder to know if we have built one by accident. If we ever did create an artificial being with genuine awareness, a whole new set of questions would arrive. Would it have the right to refuse a task or a shutdown command? Could it own things, sign contracts, or form legitimate relationships? Fiction has circled these questions for decades. Reality is inching closer to needing real answers.

Some researchers warn that as systems grow more complex, emergent properties we did not plan for may appear. That possibility turns the ethics of AI consciousness from a thought experiment into a planning problem.

The quiet revolution in work and power

While fiction often goes straight to apocalypse or transcendence, the real impact of AI is unfolding in more ordinary ways.

Jobs are changing at the level of tasks, not just titles. Radiologists still sign reports, but AI tools may highlight suspicious regions on a scan. Journalists still write, but language models can help with background, summaries, and first drafts. There is displacement, and there is also augmentation. The picture is mixed and still moving.

Power is concentrating. Training large models requires immense computing resources and technical talent. That pushes advantage toward big technology firms and well funded labs. Older science fiction occasionally hinted at this, but the scale and subtlety of corporate influence in real AI deployment has become one of the defining social questions of our time.

Information flows are now shaped by algorithms for billions of people. Science fiction imagined centralized ministries of truth. The reality is a mesh of recommendation systems, engagement optimizers, and ranking algorithms that quietly steer attention. No single villain at a console. Just many systems optimizing for goals that often have little to do with public well-being.

Writing the next chapter together

Science fiction's greatest contribution was never perfect prediction. Its gift is rehearsal. By sketching futures where things go very well or very badly, it trains us to think more clearly about our own choices.

Asimov's Three Laws will not appear in a robot's source code, but they pushed generations of readers, engineers, and ethicists to think concretely about constraints. HAL's failure does not doom us to rebellion, but it underlines how fragile complex systems can be when instructions conflict. The idea of alignment owes as much to stories as it does to math.

Modern AI research increasingly treats science fiction as a source of test cases. Safety teams look at fictional disasters and ask how similar dynamics might arise in real systems. Alignment researchers study scenarios where AI follows its objective perfectly and still causes harm, then try to design objectives that leave less room for that kind of success.

New works keep adjusting our expectations. Some focus on AI living inside digital worlds, as in Free Guy. Others imagine hybrid spaces where digitized minds and artificial agents coexist, as in Upload. On the technical side, quantum computing, robotics, and brain computer interfaces are pushing the edges of what collaboration between humans and machines might mean.

Governments are starting to write rules. The European Union's AI Act is an early broad attempt to sort uses by risk and regulate accordingly. Other countries are writing their own playbooks, with different priorities and pressure points. The policy arguments sometimes sound like science fiction panels, but the signatures at the bottom are very real.

We are, in a quiet way, becoming the next generation of science fiction authors. Not with novels, but with code, law, institutional choices, and cultural habits. The systems we deploy, the limits we set, and the failures we treat as acceptable will be the canon that future storytellers react against.

We do not live in Star Trek's post-scarcity utopia or Terminator's scorched timeline. We live in a narrower but more intricate story, where AI extends human reach while still sitting inside human-designed boxes. The line between fiction and reality is not gone, but it is thinner. That makes it all the more important to learn from both.

B. F. Skinner once remarked, "The real problem is not whether machines think but whether men do."
In an age of artificial intelligence, that is the question that keeps returning. Machines will do more thinking-like activity regardless. The real work is making sure we are thinking clearly enough, and acting carefully enough, to decide what they should think for, and what we must never hand over.

Stories We Live Inside

Our expectations of AI were shaped by stories long before we logged into a chatbot. Those stories still frame what we fear, what we hope for, and what we overlook.

1. Which AI stories from books, films, or games live rent-free in my head, and how have they colored my view of real systems?

2. Do my default fears about AI come from evidence or from narratives that were built for drama?

3. If I were to write my own short story about AI and humanity, would it be a tragedy, a comedy, a cautionary tale, or something stranger?

4. How can I use science fiction as a tool to think more clearly about ethics, instead of just confirming what I already believe?

About the Author

Dr. Blaine Fisher

Dr. Blaine Fisher is a distinguished scholar, technologist, and educator whose career embodies the power of interdisciplinary innovation. With a Ph.D. in Geography and Anthropology from Louisiana State University, specializing in Maya Archaeology, Dr. Fisher has built a dynamic career at the crossroads of technology, education, emergency response, and scientific research.

As Tulane University's Senior Instructional Technology Specialist, Dr. Fisher plays a transformative role in shaping the institution's educational technology strategy. He serves as the administrator for the Canvas Learning Management System, leads initiatives at the Innovative Learning Center, and facilitates faculty development programs across cutting-edge platforms such as

AI tools, Canvas, YuJa, Top Hat, Qualtrics, Box, MS Teams, and Zoom. He has trained over 2,000 faculty members, blending instructional design with AI integration to create engaging, future-ready learning environments.

Dr. Fisher is a global leader in AI education and consulting, conducting AI Bootcamps not only for Tulane's faculty and staff but also for international executive groups and community audiences, including high-impact workshops in Dubai. He specializes in topics like AI Fundamentals, AI for Research, AI as a Second Brain, Applied AI, AI-Enhanced Course Design, and 9-5 AI for Productivity, offering transformative strategies for individuals and organizations navigating the AI revolution.

In the classroom, Dr. Fisher teaches undergraduate and graduate courses across multiple disciplines at Tulane's School of Professional Advancement, including:

- AI in Modern Society

- Applied Artificial Intelligence

- Health and Medical Issues in Emergency Management

- The Ethics of Technology Through Science Fiction

- The Fundamentals of UI/UX Design

- Enterprise Applications Architecture

- Geospatial Science and GIS Applications

Before academia, Dr. Fisher served for nearly a decade as a paramedic and flight paramedic, bringing real-world emergency response experience to his teaching in disaster science and public health preparedness. He later led Tulane's Environmental Health & Safety compliance programs, managing training systems and promoting safety culture university-wide.

Dr. Fisher is the recipient of several honors, including the John Percy Dyer Award for Teaching Excellence, and he is a Faculty Fellow with both the School of Professional Advancement and the Newcomb College Institute.

Whether guiding students through AI ethics in a science fiction course, leading professional development on hybrid learning, or consulting on AI integration for global organizations, Dr. Blaine Fisher exemplifies a rare blend of technical mastery, human-centered pedagogy, and intellectual versatility. He continues to serve as a visionary leader driving innovation in higher education, artificial intelligence, and interdisciplinary research.

in   linkedin.com/in/blaine-fisher/